Acta Biotechnologica

Volume 6 · 1986 · Number 2

Journal of microbial, biochemical and bioanalogous technology

Akademie-Verlag
Berlin

ISSN 0138-4988
Acta Biotechnol., Berlin 6 (1986) 2, 107—206

Instructions to Authors

1. Only original papers that have not been published previously will be accepted.
   Manuscripts may be submitted in English, German or Russian (in duplicate). The name of the institute (with the full address) from which the manuscript originates should be stated below the name(s) of the author(s). The authors are responsible for the content of their contributions.

2. Original papers should not exceed 20 typewritten pages (double spaced), including references, tables and figures; short original communications may contain a maximum of six typewritten pages.

3. Each paper should be preceded by a summary in English and the title in English.

4. Latin names of species as well as passages to be printed in italics for greater emphasis should be marked by a waving line. Please use only units and symbols of the SI-system.

5. Tables may be used to shorten the text or to make it more comprehensible. They should be numbered consecutively throughout the text and be supplied with a brief heading. They should not appear in the text, but should be written on separate sheets.

6. The numbers and sizes of illustrations should be limited to a minimum, they should be numbered consecutively and be quoted on separate sheets.
   Line drawings, including graphs and diagrams, should be drawn in black ink. Half-tone illustrations should be presented as white glossy prints. Figure legends are to be typed in sequence on a separate sheet.
   The back of each sheet should bear the name(s) of the author(s).

7. References listed at the end of the contribution should contain only works quoted in the text. They should be numbered in the order in which they are first mentioned in the text.
   Please give surnames and initials of all authors, the name of the journal abbreviated according to "Chemical Abstracts — List of Periodicals", volume number, year of publication, issue number or month, first page number. Books are to be cited with full title, edition, volume number, page number, place of publication, publisher and year of publication.

8. Notes to the text may be presented as footnotes on the same page.

9. 50 offprints are free of charge.
   Additional ones may be ordered on payment.

10. The author will receive two galley proofs for correction. They are to be returned to the managing editor (DDR - 7050 Leipzig, Permoserstr. 15, Dr. Dimter) as soon as possible.

Acta Biotechnologica

Journal of microbial, biochemical and bioanalogous technology

Edited by the
Institute of Biotechnology of the Academy of Sciences of the G.D.R.; Leipzig
and by the
Institute of Technical Microbiology, Berlin
by
M. Ringpfeil, Leipzig and G. Vetterlein, Berlin

Volume 6

1986

Number 2

AKADEMIE-VERLAG · BERLIN

"Acta Biotechnologica" publishes original papers, short communications, reports and reviews from the whole field of biotechnology.
The journal is to promote the establishment of biotechnology as a new and integrated scientific field.
The field of biotechnology covers microbial technology, biochemical technology and the technology of synthesizing and applying bioanalogous reaction systems.
The technological character of the journal is guaranteed by the fact that papers on microbiology, biochemistry, chemistry and physics must clearly have technological relevance.

Terms of subscription for the journal "Acta Biotechnologica"
Orders can be sent

- **in the GDR:** to Postzeitungsvertrieb, or to the Akademie-Verlag Berlin, DDR - 1086 Berlin, Leipziger Str. 3—4; PF-Nr. 1233;
- **in the other socialist countries:** to a book-shop for foreign languages literature or to the competent news-distributing agency;
- **in the FRG and Berlin (West):** to a book-shop or to the wholesale distributing agency Kunst und Wissen, Erich Bieber, Wilhelmstr. 4—6, D-7000 Stuttgart 1;
- **in the other Western European countries:** to Kunst und Wissen, Erich Bieber GmbH, General Wille-Str. 4, CH-8002 Zürich;
- **in other countries:** to the international book- and journal-selling trade, to Buchexport, Volkseigener Außenhandelsbetrieb der DDR, DDR-7010 Leipzig, Postfach 160, or to the Akademie-Verlag Berlin, DDR-1086 Berlin, Leipziger Str. 3—4.

Acta Biotechnologica
Herausgeber: Institut für Biotechnologie der AdW
DDR-7050 Leipzig, Permoserstr. 15
(Direktor: Prof. Dr. sc. Manfred Ringpfeil) und
Institut für Technische Mikrobiologie
DDR-1017 Berlin; Alt-Stralau 62
(Direktor: Dipl. Ing. G. Vetterlein)
Verlag: Akademie-Verlag Berlin, DDR-1086 Berlin, Leipziger Straße 3—4;
Fernruf: 2236201 und 2236229; Telex-Nr.: 114420; Bank: Staatsbank der DDR, Berlin, Konto-Nr.: 6836-26-20712.
Redaktion: Dr. Lothar Dimter (Chefredakteur),
Käthe Geyler (Redakteur),
DDR-7050 Leipzig, Permoserstr. 15; Tel.: 2392255.
Veröffentlicht unter der Lizenznummer 1671 des Presseamtes beim Vorsitzenden des Ministerrates der Deutschen Demokratischen Republik.
Gesamtherstellung: VEB Druckhaus „Maxim Gorki", DDR-7400 Altenburg.
Erscheinungsweise: Die Zeitschrift „Acta Biotechnologica" erscheint jährlich in einem Band mit 4 Heften. Bezugspreis eines Bandes 120,— DM zuzüglich Versandspesen; Preis je Heft 30,— DM. Der gültige Jahresbezugspreis für die DDR ist der Postzeitungsliste zu entnehmen.
Bestellnummer dieses Heftes: 1094/6/2.

AN (EDV) 42133

03000

Acta Biotechnol. 6 (1986) 2, 109–114

Biotechnology: Health Care, Agriculture, Industry, Environment

SIKYTA, B., PAVLASOVÁ, E., STEJSKALOVÁ, E.

Czechoslovak Academy of Sciences
Institute of Microbiology, Prague
CS-142 20 Prague 4, Videnska 270

Summary

New developments in different branches of biotechnology are discussed. The production of peptide hormones, new interferons and other lymphokines by the microbial and cell cultures, and new enzyme inhibitors of microbial origin are the most important for health care and pharmacy. The main direction in research in the agriculture represents the development of the new, very effective methods of nitrogen fixation and the production of animal growth hormones by gene-manipulated microorganisms. One of the most important field of application of biotechnology is the chemical industry, c.f. microbial production of polymers and biotransformation of compounds previously produced by chemical methods (acrylamide, adipic acid, naphtalen conversion, etc.). Several novel methods of degradation of the cellulosic materials are mentioned and exploitation of biotechnology in environmental protection is also discussed.

The interest of the public in biotechnology, particularly after the recent discoveries in molecular biology and gene engineering, has centered mostly around the future impact of biotechnology on society. Biotechnology is expected to affect significantly the development of medical care, agriculture and food industries, to make more effective existing bioproductions in industry, lower the energy costs of new production processes making use of retrievable sources of raw materials and prevent further deterioration of life environment by giving rise to waste-free technologies. Biotechnological methods and bioindustrial products find their use in diverse regions: in health care and pharmacy, industry (mostly chemical), agriculture and food industry, in production of energy and in environmental control and conservation.

Health Care and Pharmacy

The crucial importance of biotechnology for medical care has been demonstrated in the past in the production of antibiotics, alkaloids, hormones, amino acids, vaccines, vitamins and other products. Elaboration of methods of gene manipulation brought a qualitative change in biotechnological methods of production of important substances, in particular human and animal proteins whose basic characteristics and effects were known but their application in therapy could not be tested owing to their poor availability. A number of peptide hormones that were previously isolated from various mammalian organs can be obtained from microbial or cell cultures to which appropriate genes have been introduced by gene manipulation techniques. The first was somatostatin, a human polypeptide cloned in *Escherichia coli*. The cloning of chemically synthesized DNA strands coding for a known amino acid sequence served to prepare insulin from *Escherichia coli*. The production of somatoliberin launched at the Centre de Recherche, France, has reached 10 g per month, the yearly demand in France being 8 g; human growth hormone is produced in *Escherichia coli* with a yield of 100 mg/l nutrient medium. Integrated Genetics (USA) produces two human fertility hormones: stimulating follicle and luteinizing hormone. Their production is not performed with genetically manipulated bacteria which do not form glycosylated products and do not excrete hormones, but with a suspension culture of cloned mouse cells into which genes governing the synthesis of these human hormones have been introduced.

Considerable hopes are placed on interferons which exhibit antiviral, antiproliferative and immunostimulating effects. Four types of interferons are known so far: α-produced by leucocytes, β-produced by fibroblasts, γ — produced by T-cells and ε-produced by epithelial cells of epidermis. With the exception of ε all can be produced in microbial cultures and in cultures of mammalian or even insect (*Drosophila*) cells after appropriate gene manipulation. It is highly probable that other types of interferons will be discovered in other cell types. Production of interferons α, β, and γ is mostly accomplished with the use of *Escherichia coli* which synthesizes 1—2 g interferon per litre culture. However, interferons are not excreted into the medium and the interest has thus shifted to *Bacillus subtilis* which produces 10-fold higher amounts of interferons and excretes it into the medium. Prokaryotic microorganisms do not synthesize interferons in the glycosylated form; as it is not clear whether glycosylated interferons are not more effective, the biosynthesis is performed with yeasts and suspension cultures of human cells. For instance, Bioferon (West Germany) has introduced the production of β-interferon in cell cultures of human fibroblasts attacked by a virus; the compound is marketed under the name Fiblaferon and is used for therapy of *Herpes zoster* infections. Gene manipulation may be used to prepare modified interferons. Cetus Corp. (USA) developed (Ser^{17}) β-interferon which exhibits a higher antiviral activity than β-interferon. Interleukin-2, the growth factor of T-cells, is a biologically active compound which, like interferons, belongs to lymphokines. It is tested for therapy of AIDS and other immune deficiencies. It is prepared by cultivating *Escherichia coli* cells obtained by cloning of genes for interleukin-2 or by cultivating lymphocytes on a pilot-plant scale. Celltech Ltd. (England) and Sankyo Co. (Japan) have developed a technology for microbial production of two human lymphokines: the factor necrotizing human tumors and interleukin-1. Both are tested in the therapy of some tumors and parasitic infections.

Human coagulation factors VIII and IX for the therapy of some types of hemophilia are also produced in submerged cultures. Factor VIII is prepared in a culture of mammalian cells while factor IX, which is therapeutically more important, in a genetically manipulated culture of *Escherichia coli*. The production has been introduced by Transgéne (France). Activator of tissue plasminogen, which activates plasmin that dissolves blood clots, can be prepared in cell cultures as well as in cultures of cloned bacteria. Which method is used is dictated by the role played by glycosylation.

An important innovation in pharmacology is the search for new enzyme inhibitors of microbial origin which should prevent enzyme degradation of antibiotics by pathogenic microorganisms, inhibit enzymes hydrolyzing saccharides and polysaccharides, inhibit proteases, and act as immunomodulators. Inhibitors of enzymes are likely to play a very important role in the development of an effective chemotherapy. Reduction of concentration of glucose in blood plasma of diabetic patients is achieved by inhibitors of intestinal invertase and amylase; the treatment prolongs the period for which saccharides and polysaccharides reside in the intestine. Apart from the well known inhibitors of proteases produced by *Streptomycetes* (leupeptine, antipain, chymostatin, elastinal, etc.) genetically manipulated microorganisms have yielded two important protease inhibitors — egline and α-1 antitrypsin. Egline is synthesized by *Escherichia coli* cells in which the genes encoding the synthesis of this inhibitor have been cloned. The synthesized amount is about 10 mg per litre. α-1 antitrypsin is produced by yeast cloned in a similar way. Both substances are tested as possible therapeutic agents for emphysema and septic shocks.

Agriculture

Products manufactured by bioindustries have contributed considerably to enhancing plant and animal production. Examples may include the production of microbial insecticides, specific antibiotics protecting plants against bacterial and fungal diseases, preparations from *Rhizobium* sp. promoting the fixation of atmospheric nitrogen in the root system of leguminous plants. Animal production has felt the impact of the industrial microbial production of essential amino acids added to fodder mixtures and production of proteins from single-cell organisms fermenting agricultural as well as industrial wastes, production of antibiotics and other bioproducts for prevention and therapy of infectious diseases of farm animals and poultry, substances stimulating their growth, and preparation regulating oestrus and the time of delivery.

New possibilities of increasing plant and animal production have appeared after introduction of the gene manipulation methods. These are represented mainly by increased fixation of atmospheric nitrogen by plants which would reduce the demand for nitrogen fertilizers. Considerable hopes were placed on the transfer of genes encoding nitrogen fixation (nif genes) from the bacterium *Klebsiella aerogenes* to plant cells. Although the transfer was successful, their expression failed to take place.

A more effective method is the fixation of atmospheric nitrogen by bacteria of the plant root system. New, recently isolated microorganisms such as *Azospirillum* (a total of 30 species) make it possible to reduce the consumption of nitrogen fertilizers in maize and clover by 25—30%. Bacteria of the genus *Rhizobium* fixing atmospheric nitrogen produce hydrogen as a byproduct; since they do not synthesize dehydrogenases, they lose 30—40% of the total energy. Recombinant technique makes it possible to introduce into these bacteria genes coding for the synthesis of dehydrogenases (hup genes) which enable these organisms to partially utilize the produced hydrogen and increase thus the efficiency of nitrogen fixation.

In northern countries considerable loses are caused by freezing of cereals and other farm crops. The accretion of ice on the plant surface was found to be caused by wild strains of bacteria colonizing the plants, especially genera *Pseudomonas* and *Erwinia*, which form proteins iniciating the formation of ice crystals from water. Protection of the plants against freezing could involve the following methods: isolation of mutants lacking this functional protein, dispensing these mutants on the surface of plants and thus replacement the wild population, or the preparation of bacteriophages lyzing the wild strain bacteria, colonization of the plants with bacteria forming no functional protein, or reduction of the adhesivity of wild bacteria to the plant surface.

Animal production will be greatly affected by economically feasible biotechnological production of growth hormones by gene-manipulated microorganisms containing genes encoding the synthesis of bovine, porcine or poultry growth hormones. The first of these methods will probably be the production of bovine growth hormones to increase milk production; experimental application of these hormones showed that a 40% increase in milk production can be safely expected. The fodder dose necessary for maintaining the basic metabolic functions, i.e. about 1/3 of the total daily dose, remains the same after hormone application but consumption of the fodder in milk production decreases. Still more important could be the growth hormones in production of pork and poultry meat. Systematic introduction of these hormones in the United States alone was calculated to require 90 tonnes of appropriate preparations with a total price of about 1 billion dollars. Experiments are also carried out with direct genetical alteration of the animals' growth. For instance, genes encoding the synthesis of the growth hormone have been transferred from rats to mouse embryas which then yielded giant mice.

Industry

One of the economically most important fields of application of biotechnology is chemical industry. Some chemical products will be produced only by biological methods, other will be obtained by combined technologies including both chemical and biological reactions. Chemical production often makes use of high pressures and temperatures and of substances causing environmental pollution; biological procedures could remove these drawbacks.

Highly topical is the production of polymers such as polyethylene and polypropylene by bacteria. During growth on acetate, *Alcaligenes eutrophus* accumulates in cells up to 80% poly-β-hydroxybutyric acid referred to the cell dry matter. Intracellular granules contain a polyester soluble in chloroform and ether, which consists of a chain of about 60 residues of poly-β-hydroxybutyrate and which can be readily polymerized. Such a polymer prepared biotechnologically has a number of advantages as compared with polyethylene, for instance better conductivity, susceptibility to biological degradation, etc.

Acrylamide has so far been produced by hydration of acrylonitrile in the presence of a catalyst such as sulphuric acid or copper, both of which have a strongly negative impact on the life environment; moreover, the production is expensive. Application of enzyme systems from bacteria of the genera *Nocardia*, *Microbacterium*, *Corynebacterium* and *Pseudomonas* makes it possible to perform the hydration by a gentle biological method at normal temperature and pressure. A combination of genes of two bacteria, *Escherichia coli* and *Pseudomonas putida*, may be used to convert naphthalene concomitantly to three products: indigo, insecticides and aspirin. Enzymes from marine animals in combination with a chelating agent may be used to degrade limestone and cement lumps to powder, which is extremely important for manufacturing high grade lime and high quality concrete. Massive and economically feasible production could probably be ensured by cloning the genes from the marine animals, encoding the appropriate enzymes, in microorganisms. Human estrogen, i.e. 17-β-estradiol, is produced by the yeast *Saccharomyces cerevisiae* as a pheromone for intercellular communication; this could also be utilized as an important source of estrogens for therapy.

The future textile industry may launch the production of new types of fabrics, especially non-woven, with novel and improved properties from the filaments of microscopic fungi, which are considerably thinner than current fibres such as rayon. These fabrics could be used for the manufacture of shoes, filter materials, medical and sanitation materials. Production of linen textiles may also be the place of application of biotechnological processes. Efforts have already been made to replace flax steeping in the fields by modern procedures such as artificial inoculation of the fibres by a specific microflora. Enzymes can now be prepared which exert a specific action, giving rise to flax fibres of exactly defined shape and chemical and physical properties.

The chapter on agriculture mentioned the negative effect of wild strains of genera *Pseudomonas* and *Erwinia* on freezing of cereals. On the other hand, these bacteria can be used to produce artificial snow of a higher quality than that of artificial snow prepared by conventional methods. This snow would permit skiing at temperatures higher by 3—4°C. Water is inoculated by lyophilized bacteria catalyzing the formation of ice crystals which have the same shape as natural snowflakes. The quality of the snow can be regulated by changing the proportion of bacteria to water; the cost of production of artificial snow would drop by 1/3.

A highly interesting and promising field is molecular electronics which deals with the construction of integrated circuits with a high degree of integration from organic molecules (biochips) in which individual elements, i.e. individual molecules, will be about 50 nm long. These biochips will be based on protein molecules, other molecules bound to them serving as conductors, switches and other elements. These could include porphyrins (hemes and chlorophyll) which form planar arrays of molecules and contain metal ions, and also monoclonar antibodies which can be prepared with high purity and specificity. Biochips will be able to process 10^3 or even 10^6 times more bits of information than the currently used silicon microchips; this will cause revolution in the conception and construction of computers. Mass production of new molecular electronic circuits will reduce production costs of, e.g. personal computers with extensive memories. These will then be used for controlling the movements and functions of artificial limbs and organs, in construction of implantable minicomputers for the blind and the deaf, improvement of memory and control of a new generation of robots.

Energy and raw material sources and environmental protection

Plant materials containing cellulose are considered to be an important source of energy for biotechnological processes since they are renewable. The current procedures of degradation of these materials, whether chemical or biological (enzymic), are economically disadvantageous. Further possibilities are therefore sought for obtaining cellulose and hydrolysing it to saccharides from complexes containing cellulose, hemicellulose and lignin. A perspective technique could be the processing of plant biomass by a concomitant action of hydrogen peroxide and sodium hydroxide; after several hours at room temperature the biomass yields about 50% by weight of lignin and hemicellulose. Insoluble cellulose is dried. Lignin can also be removed by a 1-hour treatment with acidified mixture of butanol and water at 170°C. The ensuing products can be easily separated since lignin accumulates in butanol, hemicellulose in the aqueous phase and cellulose remains in the solid phase. Lignin can be conveniently used for sulphonations, furfurol formed from hemicellulose has application in crude oil extraction, viscose prepared from cellulose finds its use as car paint or varnish. Lignin can be removed from plant material also by microbial technology using enzymes of the fungus *Phanerochaete chrysosporium* and bacteria of the genus *Arthrobacter*.

Another raw material suitable for the production of special chemicals is cutine from the waxy cuticula of the outer cell wall of leaves, stems and fruits of plants. Cutine is a polymer which can be degraded to monomers by cutinase from the fungus *Fusarium*. In view of the possible industrial production, the gene regulating the synthesis of cutinase has been transferred from *Fusarium* to *Escherichia coli*.

Biotechnological methods of liquidation of solid, liquid or gaseous wastes represent the most efficient and cheapest way of coping with these wastes. Anaerobic procedures of processing of (especially agricultural) wastes to biogas have gained wide publicity. Mixed cultures of microorganisms in waste waters assimilate nitrates and phosphates but the process is incomplete. When the appropriate enzymes are induced by adding nitrates and phosphates to the growing population, these substances can be removed even in trace amounts. Some molecules produced by microbes were found to bind with almost 100% efficiency heavy metals such as mercury, cobalt, vanadium, cesium, strontium, etc. These could be used, e.g., for trapping radiocobalt in nuclear power stations and other metals in metallurgy. The method could be used also in pharmacy or in other branches where stress is laid on removal of heavy metals, e.g. from kitchen salt. Microorganisms are capable of binding heavy metals not only on their surface but accumulate them inside the cells in amounts up to 60% of cell dry matter. A source of problems in waste waters are phenols and aromatic amines which can be removed by hydrogen peroxide in combination with microbial peroxidase; these substances are polymerized to large molecules which precipitate easily. Within 3 hours up to 99% of these substances can be removed by precipitation.

An important facet of environmental protection would be the replacement of sodium chloride in antifreeze road sprays by other substances which are not corrosive and can be biologically degraded. Attention has been directed to, e.g. biotechnological production of magnesium calcium acetate from waste materials by bacteria such as *Clostridium thermoaceticum*. Even though the costs would be 7—8 times higher than with sodium chloride, they would be compensated by elimination of corrosion.

Among gaseous wastes the most hazardous for environment is sulphur dioxide, carbon monooxide and hydrogen sulphide. Microorganisms can assimilate these gases, with the exception of sulphur dioxide, and convert them into useful products. The anaerobic bacterium *Chlorobium thiosulphatophilum* under photoautotrophic conditions assimilates hydrogen sulphide and carbon dioxide, excretes sulphur and converts the carbon to polyglucose. In the dark it converts the polyglucose to three organic acids with acetic acid as the main component. *Pseudomonas carboxydovorans* and various strains of the genus *Bacillus* utilize carbon monooxide as the only carbon source and could thus be used for biosynthesis of biomass and enzymes.

Directed selection and gene manipulation can be used to select hyperproducing microorganisms yielding high amounts of enzymes, enzymes with altered substrate specificity and completely new enzymes capable of degrading toxic substances, xenobiotics, plastics, etc. This would facilitate the development of new efficient methods of degradation of toxic substances in the environment and promote the production of chemicals by biotechnological methods.

Received April 13, 1985

Acta Biotechnol. 6 (1986) 2, 115—121

Purification and Characterization of two Extracellular β-Glucosidases from Trichoderma viride ITCC 1433

WILHELM, M. [1], SAHM, H.[2]

[1] Institut für Mikrobiologie
Universität Düsseldorf
D-4000 Düsseldorf 1, FRG
[2] Institut für Biotechnologie
Kernforschungsanlage Jülich
Postfach 1913, D-5170 Jülich, FRG

Summary

T. viride ITCC 1433 synthesizes a two component system for the hydrolysis of cellobiose and cello-oligodextrins. 80% of the total activity are solubilized during growth. The large protein (A), mol. weight 98000 d, is glycosylated and slightly acidic (pH = 6.1). The smaller protein (B), mol. weight 70000 d, is unglycosylated and neutral (pH = 7.2). Both proteins form a two-step system where β-glucosidase A is active at low substrate concentrations ($K_M = 2.3 \times 10^{-4}$ M cellobiose) while β-glucosidase B covers the range of 10-fold higher cellobiose concentrations ($K_M = 1.8 \times 10^{-3}$ M). The enzymes are fairly stable with a residual activity of 70% at 50 °C after 24 h.

Introduction

Recently the interest in enzymatic cellulose hydrolysis has become very strong because of the world wide interest in utilizing renewable resources of biomass as a source of chemicals and liquid fuels. A large number of cellulolytic microorganisms have been studied as potential sources of cellulases for use in the saccharification of cellulosic materials and it was found that the cellulolytic enzymes of *Trichoderma* species are the most active for hydrolysis of crystalline cellulose. However, the level of β-glucosidase is very low in these enzyme preparations. β-Glucosidase is not only important for the production of glucose from cellobiose and small cellodextrins but also plays a crucial role in practical saccharification by removing the cellobiose inhibition of cellulases. Therefore the rate and extent of saccharification of cellulose could be increased by the addition of β-glucosidase from *Aspergillus phoenicis* [1].

In a preceding paper it was demonstrated that *Trichoderma viride* ITCC 1433 secreted in addition to high amounts of β-glucosidase high levels of cellulases during growth on cellulose. The purification and characterization of two β-glucosidases from this *Trichoderma* strain are presented in this report.

Materials and Methods

Materials

Cellulose powder MN 300 and Avicel were purchased from MACHAREY and NAGEL (Düren, F.G.R.). Desoxynojirimycin was generously donated by Prof. Dr. W. FROMMER (Bayer, Leverkusen, F.G.R.). Glucose-6-phosphate dehydrogenase and hexokinase were obtained from BOEHRINGER

(Mannheim, F.G.R.) and Biogel P-200 from Bio-Rad Laboratories (München, F.G.R.). p-Nitrophenyl-β-D-glucopyranoside and Servalyt ampholines were purchased from SERVA (Heidelberg, F.G.R.). Other chemicals used were of chemically pure quality.

Organism and Cultivation

The organism used in this study was *Trichoderma viride* ITCC 1433, a mesophilic cellulolytic fungus which was isolated by GUPTA et al. [2]. It was grown in 500 ml conical flasks containing 100 ml of MANDELS medium [3] consisting of different mineral salts, 0.1% (w/v) proteose peptone, 0.2% tween 80 and 1% glucose, cellobiose or cellulose as carbon and energy source. The pH of the medium was adjusted to 6.0. Spores were used to inoculate the medium (10^6 spores/flask) and incubation was carried out at 27 °C on a rotory shaker (100 Upm). pH was kept above 3.0 by the addition of 2 N NaOH.

Enzyme Assays

The β-glucosidase activity was determined with p-nitrophenyl-β-D-glucopyranoside (pnpg) or cellobiose as substrate [4]. A 0.9 ml aliquot of a 10 mM pnpg solution in 50 mM sodium acetate buffer, pH 4.8, was heated to 50 °C, then 0.1 ml of the enzyme solution was added to the substrate. After 10 min incubation at 50 °C 2 ml of 1 M sodium carbonate was added to stop the reaction and the p-nitrophenol released was determined by extinction measurements at 405 nm.
When using the natural substrate, assays typically contained 10 mM cellobiose, 50 mM sodium acetate buffer, pH 4.8, and enzyme at an appropriate dilution, in a final volume of 1.0 ml. After 10 min incubation at 50 °C, enzyme reactions were stopped by boiling for 5 min. Glucose release was determined enzymatically [5] or with a Biotronik Sugar Analyzer [6]. One unit of β-glucosidase activity is defined as that amount of enzyme protein which hydrolyzes 1 μmol of pnpg or cellobiose per minute under the assay conditions.

Purification of the β-glucosidases

Unless otherwise stated, all the following procedures were carried out at 4 °C.

Step 1. Crude Enzyme. T. viride cells grown for 5 days (600 ml culture) on cellulose were removed by centrifugation. The supernatant was concentrated by ultrafiltration (ultrafilter PM 10 from Amicon) to one-sixtieth of the original volume and dialyzed for 16 h against two changes of 10 mM sodium acetate buffer (pH 4.8). The resultant crude enzyme was further purified by chromatofocusing.

Step 2. Chromatofocusing. The concentrated supernatant from the previous step (9 ml) was applied on to a polybuffer ion exchanger column (PBE 94, Pharmacia) equilibrated with 25 mM ethanolamin — acetic acid buffer pH 9.0. Elution was carried out with amphoteric polybuffer (Polybuffer 96, Pharmacia) pH 6.0, thus creating a pH gradient between 9 and 6 and the proteins were eluted according to their isoelectric points. The flow rate was 14 ml/h and 1 ml fractions were collected. The fractions with high enzyme activity were combined and concentrated by ultrafiltration.

Step 3. Gel filtration. The two concentrated enzyme solutions were purified further by applying the two portions on a column (5 × 100 cm) packed with Bio Gel P 200 equilibrated with 100 mM sodium phosphate buffer (pH 6.8) containing 100 mM NaCl and 0.02% sodium azide. The column was eluted with the same buffer at a flow rate of 5 ml/h and 2 ml fractions were collected. The fractions showing β-glucosidase activity were dialyzed against 50 mM sodium acetate buffer (pH 4.8) with 0.02% sodium azide, stored at 4 °C and used as enzyme source for further studies.

Protein concentration was determined by the methods of BENSADOUN and WEINSTEIN [7].

Polyacrylamide Gel Electrophoresis

Analytical polyacrylamide gel electrophoresis was performed in 6.8% gels. The separation gels contained in a final concentration 0.375 M Tris-HCl buffer, pH 8.9 according to MAURER [8]. The gels were stained with Coomassie brillant blue according to FAIRBANKS et al. [9]. For estimation of the molecular weights of the enzymes sodium dodecylsulfate gel electrophoresis was carried out as described by Laemmli [10]. The following reference proteins were used: RNA polymerase from *E. coli* and its subunits (165000, 155000, 39000), phosphorylase B (94000), bovine serum albumin (68000), ovalbumin (43000). Gels to be stained for carbohydrates were fixed and stained by the periodic acid-Schiff base technique described by Maurer [8].

Estimation of Molecular Weight by Gel Filtration

The molecular weights of the two β-glucosidases were determined according to the method of Andrews [11] by gel filtration on Bio Gel P 200. The column was equilibrated with 0.1 M sodium phosphate buffer, pH 6.8. The enzymes and the calibration proteins: catalase (232000), bovine serum albumin, dimer (136000), ovalbumin (43000), chymotrysinogen (25000) were eluted with the same buffer.

Isoelectric Focusing

In order to determine the isoelectric points of the purified proteins, analytical isoelectric focusing was performed in Servalyt polyacrylamide gel slabs with a pH gradient, ranged from pH 6 to 10. The focusing was run at 2.25 Watts for 2.5 h at 8 °C. The following reference proteins were used: ovalbumin (pH: 5.9), horse myoglobin (pH 7.3), vale myoglobin (pH 8.3), ribonuclease (pH 9.5), cytochrome C (pH 10.7). Proteins in the gels were stained with Coomassie brillant blue.

Results and Discussion

Synthesis of β-glucosidase on Different Carbon Sources

During the active growth phase of *Trichoderma viride* ITCC-1433 on cellulose, the β-glucosidase activity rises to 5.8 U/ml measured with the substrate pnpg or 2.3 U/ml determined with cellobiose. This is about the 7-fold amount which will be obtained with

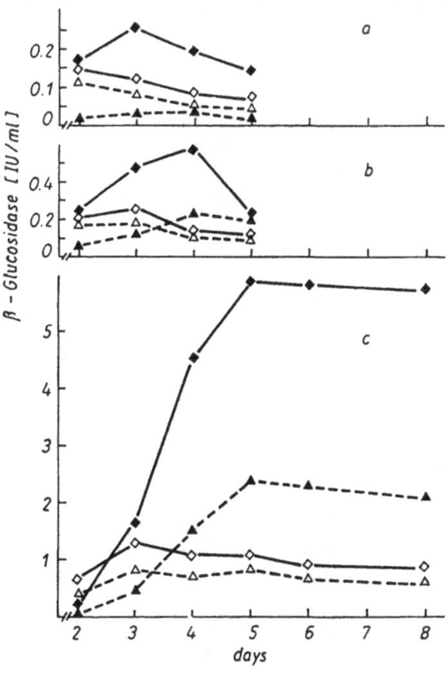

Fig. 1. The time course of the β-glucosidase formation with a) glucose; b) cellobiose; c) cellulose as substrate is shown.
Solid marks indicate soluble activity (rhombs: activity determined with pnpg as substrate; triangles: activity determined with cellobiose).

Trichoderma reesei MCG 77. The enzyme production on cellobiose is ten times less than on cellulose and just two times higher than on glucose. This fits well with the results reported by VAHERI et al. [12] and EVELEIGH and MONTENECOURT [13], and is due to the catabolite repression of glucose which is produced by the hydrolysis of cellobiose. As shown in Fig. 1 most of the β-glucosidase activity is excreted into the culture medium.

Purification of the β-glucosidases

In order to determine the number of the β-glucosidases and their isoelectric points, the broth of a cellulose culture was subjected to preparative isoelectric focusing with a gradient range from pH 2 to pH 11. By testing the β-glucosidase activity of the separated proteins two active bands were detected. Chromatofocusing was chosen as the first separation step (Fig. 2). The indicated fractions A and B were pooled, concentrated and subjected separately to further purification with a Biogel P200 column. After this second purification step two protein fractions with β-glucosidase activities were obtained and both proteins were gelelectrophoretically homogeneous (Fig. 3). The purification procedure is summarized in Table 1.

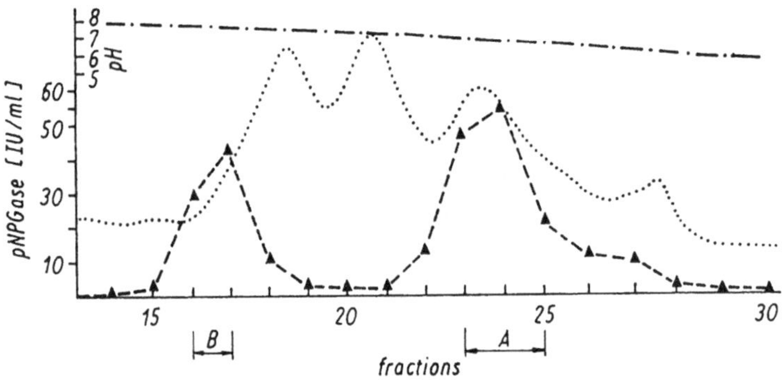

Fig. 2. Chromatofocusing of crude enzyme preparation from a *T. viride* cellulose culture. 200 mg of protein were loaded onto the column.
Dashed and dotted line: pH gradient. Dotted line: Absorbance at 280 nm. Dashed line, triangles: p-nitrophenyl-β-D-glucopyranosidase activity.

Table 1. Purification of two β-glucosidases from the culture supernatant of *Trichoderma viride* ITCC-1433

	Volume [ml]	Total Protein [mg]	Total activity (C-Base) [U]	Specific activity [U/mg]	Recovery yield [%]	Purification -fold
Culture supernatant	600	618	1440	2.3	100	1
Ultrafiltrate	9	198	698	3.5	49	1.5
Chromatofocusing	50	—	563	—	39	—
Biogel P 200						
Protein A	24	4	260	65	18	28
Protein B	25	1.4	19	14	1.3	6

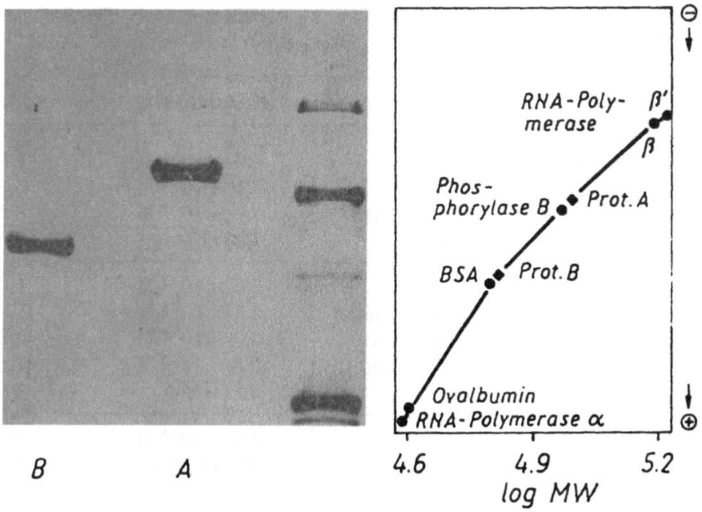

Fig. 3. SDS gel electrophoresis of the two purified β-glucosidases A and B. The calibration curve given in the right half is derived from the marker proteins shown on the gel. 15 μg of protein A and B, respectively, were loaded onto the gel.

Characterization of the β-glucosidases

The molecular weights of the enzymes were determined by gel chromatography on Biogel P200 to be 130000 (A) and 62000 (B). When the enzymes were subjected to gel electrophoresis in the presence of sodium dodecyl sulfate each enzyme showed a single protein band, the mol weights being 98000 (A) and 70000 (B), respectively. From these results we assume that both proteins are composed of a single polypeptide chain, with a molecular weight similar to the β-glucosidases described from *Trichoderma reesei* [14, 15]. The deviation of the molecular weight of protein A in SDS electrophoresis is, to our opinion, explained by the glycosylation of the protein (see below). The isoelectric points of the β-glucosidases were determined by analytical isoelectric focusing to be 6.1 for A and 7.2 for B. These data are slightly different from those reported for the β-glucosidases from *Trichoderma reesei* or *Trichoderma koningii*. They have a pH-range from 5.74 to 5.85 [16, 17].

As shown in Table 2 both β-glucosidases of *Trichoderma viride* are not only able to hydrolyse cellobiose or p-nitrophenylglucoside but also some cellodextrines as for example cellopentaose. This sugar oligomer is decomposed by an exocleavage since the first main reaction products are cellotetraose and glucose. Both K_m-values are extraordinarily low in comparison to the data obtained for the enzymes of *Trichoderma reesei* [14, 18]. It is very surprising that both enzymes had lower K_m-values and higher V_{max}-data for cellopentaoses than for cellobiose; it seems that this two β-glucosidases are also responsible for the decomposition of cello-oligomers formed from cellulose by endocellulases. Those two β-glucosidases are highly sensitive to competitive inhibition by glucose and also by the glucose structure analoge desoxynojirimycin. As shown in Fig. 4 also fructose inhibits these enzymes in a noncompetitive manner and to a lower extend. This may be of interest for the technological application of these enzymes since the end-product inhibition of glucose can be decreased by enzymatic isomerisation into fructose.

As extracellular fungal enzymes are frequently glycoproteins purified β-glucosidases were tested for carbohydrate content by selective staining techniques. The enzyme solutions

Table 2. Kinetic properties of the two β-glucosidases from *Trichoderma viride* ITCC-1433

Substrates		Enzyme A	Enzyme B
p-Nitrophenylglucosid	K_M:	4.9×10^{-5} M	1.2×10^{-4} M
	K_i (Gluc):	3.4×10^{-4} M	5.5×10^{-4} M
	K_i (Des):	4.0×10^{-7} M	8.6×10^{-7} M
	V_{max}:	125 IU/mg	270 IU/mg
Cellobiose	K_M:	2.3×10^{-4} M	1.8×10^{-3} M
	K_i (Gluc):	2.5×10^{-4} M	1.3×10^{-4} M
	K_i (Des):	3.0×10^{-7} M	8.3×10^{-7} M
	K_{ii} (Fruc):	1.5×10^{-2} M	1.6×10^{-2} M
	V_{max}:	90 IU/mg	67 IU/mg
Cellopentaose	K_M:	8.0×10^{-5} M	9.0×10^{-5} M
	V_{max}:	100 IU/mg	105 IU/mg

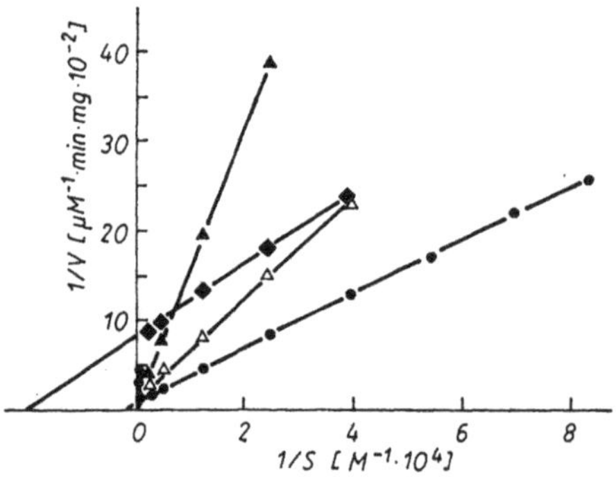

Fig. 4. LINEWEAVER-BURK plot of the cellobiase reaction of protein A.
Dotted line: various concentrations of cellobiose without inhibitors. Open triangles: same as above plus 10^{-4} M glucose (competitive inhibition). Closed triangles: same as above plus 10^{-6} M desoxynojirimycin (competitive inhibition). Rhombs: same as above plus 10^{-1} M fructose (noncompetitive inhibition). The dotted line also indicates the phenomenon of substrate inhibition at high concentrations.

Table 3. Carbohydrate identification of the β-glucosidase A of *Trichoderma viride* ITCC-1433

	Con A	SBA	WGA	UEA	PNA	RCA
Specificity	-D-mannosyl -D-glucosyl	N-acetyl-galactos-aminyl	β-N-acetyl-glucosaminyl	L-fucosyl	β-D-gal-1.3--D-galNac β-D-galactosyl	D-galactosyl
Remaining activity [%]						
Protein: A	0	78	78	0	71	70

Abbreviations: Con A (Concanavalin A); SBY (Soybean Agglutinine); WGA (Wheat Germ Agglutinine); UEA I (Ulex Europaeus Agglutinine I); PNA (Peanut Agglutinine); RCA I (Ricinus Commun's Agglutinine I).

were subjected to polyacrylamide gel electrophoresis and finally the gels were stained with SCHIFF's reagent [8]. By this way it could be demonstrated that protein A contains carbohydrates while the protein B showed no colour reaction. In order to identify the carbohydrates bound to protein A the enzyme was incubated with different agarose bound lectines (Table 3). These aggregation studies indicate that the β-glucosidase (A) contains the sugars fucose and mannose or glucose which are the most common carbohydrates in the fungal cell wall [19].

Final conclusions

Trichoderma viride ITCC 1433 performs an 80% release of the total amount of cellobiase activity in shake flask culture and has therefore proven to be a good source of cellobiases. Apparently the cellobiose utilizing system consists of two different enzymes which differ in molecular weight, pH, affinity to their substrate and reaction velocity. Enzyme A covers the low substrate concentration range, whereas enzyme B is active at 10-fold higher substrate concentrations. This two step mechanism is maintained at temperatures up to 50°C, where the system shows a reasonable stability of 70% remaining activity after 24 hours. To our opinion an increase of stability can be achieved by immobilization of the proteins. A mentionable disadvantage is the sensitivity of both cellobiases to competitive glucose inhibition. A possibility to circumvent this disadvantage lies in the conversion of glucose to fructose since latter compound has proven to be 100-fold less effective in the cellobiase inhibition. On the other hand the very high sensitivity against desoxynojirimycin, a compound which cannot be metabolized by *Trichoderma*, offers the possibility to select for a cellobiase system with a lower glucose sensitivity after mutagenization. The cellobiase system of *Trichoderma viride* ITCC 1433 can supply the cellulase system of other fungi to improve the cellfree cellulose saccharification.

Received April 18, 1985

Reference

[1] STERNBERG, D., VIJAYAKUMAR, P., REESE, E. T.: Can. J. Microbiol. **23** (1977), 139.
[2] GUPTA, J. K., DAS, N. B., GUPTA, Y. P.: Agric. Biol. Chem. **36** (1972), 1961.
[3] STERNBERG, D.: Biotechnol. Bioeng. Symp. **6** (1976), 35.
[4] BERGHEM, L. E., PETTERSON, L. G.: Eur. J. Biochem. **46** (1974), 295.
[5] BERGMEYER, H.-U.: Methoden der Enzymatischen Analyse, Bd. 2. Weinheim: Verlag Chemie, 1974. 2. Aufl., p. 1241.
[6] SINNER, M., SIMATUPANG, M. H., DIETRICHS, H. H.: Wood Science Technol. **9** (1975), 307.
[7] BENSADOUN, A., WEINSTEIN, D.: Anal. Biochem. **70** (1976), 241.
[8] MAURER, H. R.: Disc Electrophoresis and Related Techniques of Polyacrylamid Gel Electrophoresis. Berlin, New York: Walter de Gruyter, 1971. P. 1 and p. 76.
[9] FAIRBANKS, G., STECK, L., WALLACH, D. F. H.: Biochemistry **10** (1971), 2606.
[10] LAEMMLI, U. K.: Nature **277** (1970), 680.
[11] ANDREWS, P.: Biochem. J. **91** (1964), 222.
[12] VAHERI, M. P., VAHERI, M. E., KAUPINNEN, V. S.: Eur. J. Appl. Microbiol. Biotechnol. **8** (1979), 73.
[13] EVELEIGHT, D. E., MONTENECOURT, B. S.: Adv. Appl. Microbiol. **25** (1978), 57.
[14] GONG, C.-S., LADISCH, M. R., TSAO, G. T.: Biotechnol. Bioeng. **19** (1977), 958.
[15] INGLIN, M., FEINBERG, B. A., LOEWENBERG, J. R.: Biochem. J. **185** (1980), 515.
[16] PETTERSSON, L. G.: Proc. Symp. Enz. Hydrol. Cellulose, Aulanko, Finland, 1975, p. 255.
[17] WOOD, T. M.: Biotechnol. Bioeng. Symp. **5** (1975), 111.
[18] BERGHEM, L. E., PETTERSSON, G.: Eur. J. Biochem. **46** (1974), 295.
[19] AINSWORTH, G. C., SUSSMANN, A.: The Fungi. New York, London: Academic Press, 1965.

Acta Biotechnol. 6 (1986) 2, 122

Book Reviews

William R. STROHL, Olli H. TUOVINEN

Microbial Chemoautotrophy

Columbus: Ohio State University Press, 1984
351 pp., 89 fig., 36 tab., $ 32.50

Das Buch ist das Ergebnis eines Kolloquiums im Oktober 1982 an der Ohio State University. Auf 341 Seiten erfährt der Leser von 26 namhaften (fast ausschließlich amerikanischen) Wissenschaftlern in 19 Beiträgen Neues, Zusammenfassendes, aber auch in die Zukunft Weisendes über die Biologie, Chemie und Bioenergetik der chemolithotrophen Bakterien. Es handelt sich um eine sehr moderne, informative Darstellung, beginnend mit einer kurzen Einführung, in der Patrick R. DUGAN auch zur Begriffsbildung Stellung nimmt.
Chemolithotrophe sind naturwissenschaftlich außerordentlich interessant, mehr und mehr gewinnen sie jedoch praktische Bedeutung insofern, als ihre Leistungen technisch genutzt werden sollen. Auf diese Bedeutung hinzuweisen, war ein Anliegen des Kolloquiums. Die Beiträge "Desulfurization of Coal ..." von P. R. DUGAN am Anfang und "Research from an Industrial Viewpoint ..." von R. RADMER am Ende des Buches belegen dies. In Stichworten weiteres aus dem Inhalt: Oxidation von Eisen (COX & BRAND) und Nichteisenmetallen (DISPIRITO & TUOVINEN), von Nitrit (COBLEY), Ammonium (HOOPER), Kohlenmonoxid (PARK & HEGEMAN), Sulfat-Reduktion (PECK), Methanogenese (SPROTT, JARRELL, HOOK u. a.), Mixotrophie (MATIN, STROHL, SCHMIDT, GIBSON); und selbstverständlich findet man auch einen Beitrag über die Aufrechterhaltung des pH-Gradienten bei Acidophilen und Alkali(no)philen (J. G. COBLEY).
Dieses Buch kann wärmstens empfohlen werden, Biowissenschaftlern, weil sie mit anwendungsorientierten Aspekten konfrontiert werden, Biotechnologen, weil sie sich rasch mit den naturwissenschaftlichen Grundlagen industriell nutzbarer Leistungen von Mikroorganismen vertraut machen können.

W. BABEL

P. v. SENGBUSCH

Einführung in die Allgemeine Biologie

Berlin, Heidelberg, New York, Tokyo: Springer-Verlag, 1985. 526 S., 328 Abb., 59 DM

Niemand wird von einem Lehrbuch der Allgemeinen Biologie erwarten, daß es in der Erfassung auch nur der wichtigsten biologischen Erscheinungen vollständig ist. Was aber heute von einem solchen Buch erhofft wird, ist die Darstellung der Biologie als exakte Naturwissenschaft, theoretisch konsistent, unabhängig von musealer Vielfalt und extrapolierbar über das Naturgegebene hinaus bis hin zu aktuellen biologischen Technologien. Diesen anspruchsvollen Zuschnitt hat das vorliegende Buch.
Das Prinzipielle und das Methodische stehen im Vordergrund, so wird Dichte und Überschaubarkeit erreicht. Das Lebendige ist dargestellt in seinen drei Organisationsebenen: Zelle, Gewebe, Gesellschaften. Das Verständnis für die Übergänge und die steigenden Komplikationsgrade von abiotisch zu biotisch, von Ebene zu Ebene wird schonend entwickelt. Molekularbiologie, Gentechnik, Signalverwertung, Ökologie, Evolution, menschliche Produktion ordnen sich zu einer notwendigen Einheit.
Das Besondere dieser Einführung ist weniger die Beschreibung des Lebens, sondern vielmehr die Anleitung zum produktiven Umgang mit ihm. Wie wird diese aktivierende Wirkung erreicht? Der Autor versteht es, die Dinge einfach und klar, mitunter humorvoll zugespitzt, zu benennen. Er zeichnet den Weg der Erkenntnis nach, zitiert die großen Wegbereiter, fragt, entnimmt dem Alltag beweiskräftiges Material und gibt methodische Grundlagen. Der Text ist bei aller Dichte niemals strapaziös, kein Monolog. Die vorzügliche Ausstattung ist den Stilmitteln des Autors adäquat. Dieses Buch vermittelt nicht nur Wissen — es prägt biologische Denkweisen aus. Man kann es Biologie-Lehrern und Studenten ebenso empfehlen wie erfahrenen Spezialisten oder Vertretern benachbarter Disziplinen, die an einer biologischen Gesamtsicht interessiert sind.

K. SATTLER

Acta Biotechnol. 6 (1986) 2, 123—128

Gas Balance Method for Testing of Microbial Growth Efficiency after Carbon Substrate Pulse

BABURIN, L. A., SHVINKA, I. E., RUKLISHA, M. P., VIESTURS, U. E.

Laboratory of Bioengineering
August Kirchenstein Institute of Microbiology
Latvian SSR Academy of Sciences
Kleisti, Riga 226067, Latvian S.S.R, U.S.S.R.

Summary

The present paper deals with the analysis of the amount of oxygen utilized for oxidation of a small dose of carbon substrate in carbon limited *Brevibacterium flavum* culture. The ratio of the measured oxygen consumption (m_{O_2}) to the amount of added carbon substrate (m_S) gives a stoichiometric coefficient of the biological oxidation equation. A linear relationship between m_{O_2} and m_S was observed. To compare the efficiency of different carbon substrate utilization there has been introduced a normalized value $\beta = m_{O_2}^{exp}/m_{O_2}^{theor}$. There exists a simple relationship between β and the thermodynamical growth efficiency η. The theoretical considerations are proved by experimental results with β, η and $Y_{x/s}$ in a chemostat culture at various medium flow rates.

Introduction

The efficiency of microbial growth (η) is expressed as a part of chemical energy of the substrate, transferred into the biomass [1, 2]. It is defined as a normalized ratio [1—3]:

$$\eta = Y^{exp}/Y^{theor}.$$

The theoretical yield of biomass on an organic substrate can be calculated from the elemental and energy balance equation of microbial growth, using macroscopic principles [1, 2]. The experimental biomass yield is measured experimentally as:

$$Y_{x/s}^{exp} = \Delta X/\Delta S; \quad \text{or} \quad Y_{x/s}^{exp} = \mu/q_s \quad \text{(under steady-state conditions)}.$$

Microbial growth is a relatively slow process. ΔX increment is detectable during approximately an hour. ΔS is measured mainly with the help of biochemical analysis. Therefore $Y_{x/s}$ is usually evaluated as an integral parameter and its application in an express control and monitoring of the process is limited.
The pulse method is used for a qualitative testing of culture growth conditions. After the addition of a small quantity of growth-limiting substrate the culture responds with temporary changes in pH and pO_2 [4, 5] or respiratory rate [6, 7]. Gas balance method is used for the analysis of different chemostat culture responses [6] and for the optimization studies after the substrate pulse [6, 8]. The combination of gas balance method with the pulse technique allows to obtain the information on the efficiency of different carbon source utilization in physiological environment [7].

In the present work we determine the quantity of oxygen consumed by *Brevibacterium flavum* cells for the oxidation of small doses of different carbon substrates in a C-limited chemostat. The quantitative information obtained is discussed in terms of the virtual efficiency of carbon substrate utilization.

Materials and Methods

Experiments were carried out with a *Brevibacterium flavum* culture from the collection of August Kirchenstein Institute of Microbiology, Latvian SSR Academy of Sciences. We used a medium containing the following (g/l): sucrose — 25, cornsteep extract — 10, $(NH_4)_2SO_4$ — 15, K_2HPO_4 and KH_2PO_4 — 0.5 each.
Cultivation was carried out in a laboratory scale fermenter with a system for the measurement of the heat evolution rate [6]. Steady-state conditions in the chemostat were maintained by a continuous feed of the medium and removal of the culture fluid by peristaltic pumps S-32 (Poland), the volume being constant. Respiration intensity of the culture (Q_{O_2}) and Q_{CO_2}) was measured with the help of gas balance method, using the gas-analyzing system PGA-2 (designed at A. Kirchenstein Institute of Microbiology by the authors of the present article, made on the Experimental Electro-Mechanical Plant of the Physico-Energetical Institute, Latvian SSR Academy of Sciences). Establishing of the steady-state of the culture, upon medium flow, took 6 hours. Then there followed the carbon substate pulse into the culture fluid.
The total amount of oxygen consumed after the substrate addition was determined by integrating the response — changes in the rate of oxygen consumption by the culture (Fig. 1):

$$m_{O_2} = q_{nc} \int_0^{\tau} (C_{O_2}^{st} - C_{O_2}(t))\, dt \qquad (\text{mol } O_2)$$

where: $C_{O_2}^{st}$ — steady-state value of O_2 concentration in the outlet gas (air);
$C_{O_2}(t)$ — changes of O_2 concentration in the outlet gas after addition of substrate into culture fluid;
q_{nc} — air flow rate through fermenter, under normal conditions;
τ — time of response.

Concentration of biomass (X) was determined spectrophotometrically at $\lambda = 650$ nm and expressed in units of dry weight.

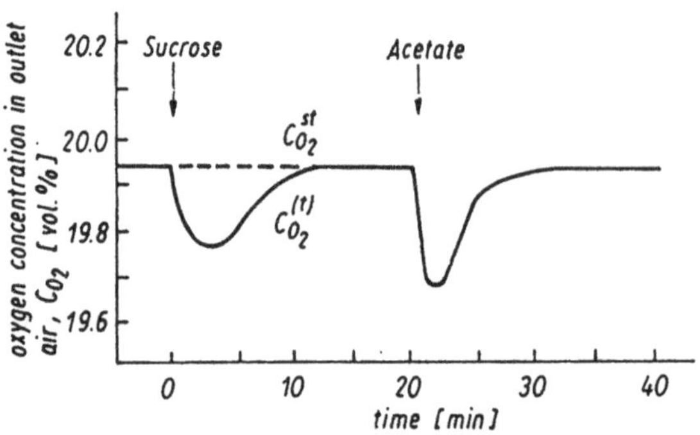

Fig. 1. Changes of oxygen concentiation in the outlet air after the carbon substrate pulses.

Residual sucrose concentration was determined according to the raection with 1,3-dinitrosalicylic acid. Biomass yield in a continuous culture ($Y_{x/s}$) was expressed as μ/q_s (g/g). A more detailed description of the methods is given in [6].

Results

Small doses of carbon substrates in *Brevibacterium flavum* cause temporary changes in the rates of oxygen consumption, CO_2 evolution and heat generation [6, 7]. In carbon-limited chemostat a carbon substrate pulse causes an increase of Q_{O_2}, which gives transient measurable changes in oxygen concentration in the outlet air (Fig. 1). During 10—15 min (it depends on the substrate dose) the Q_{O_2} value and corresponding O_2 concentration in the outlet air resume their basic level. CO_2 evolution and heat generation give similar responses (not shown). There are no detectable changes in the residual sucrose concentration in the medium. Therefore it seems, that the pulse of the carbon substrate added is completely utilized by cells.

The relationship between oxygen uptake and the quantity of the added C-substrate is shown in Figure 2. The effect of two substrates was studied, that of sucrose (growth substrate) and acetate (auxiliary substrate). For both C-substrates there is a linear

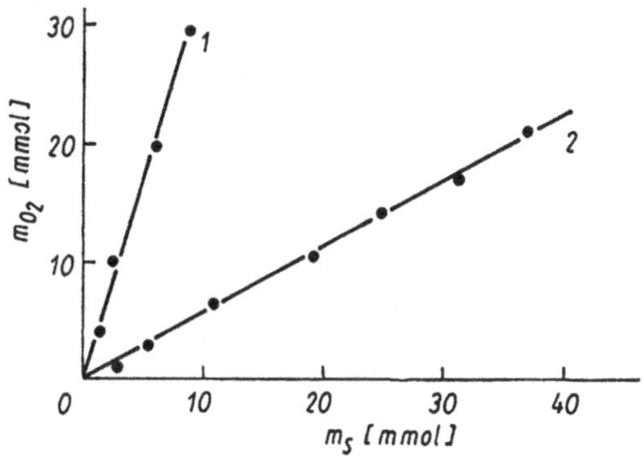

Fig. 2. The relationship between the amount of oxygen consumed (m_{O_2}) and C-substrate added (m_s).
1 — sucrose, 2 — acetate
Brevibacterium flavum chemostat culture, $D = 0.1\ h^{-1}$

relationship between m_{O_2} and m_s. The slope m_{O_2}/m_s gives the stoichiometric coefficient B for the growth equation:

$$A(CH_2O)_n + B(O_2) + D(NH_3) \rightarrow Y(CH_{1.63}O_{0.50}N_{0.13}) + (An\text{-}Y)CO_2 + T(H_2O) \qquad (1)$$

where: $CH_{1.63}O_{0.50}N_{0.13}$ — is the elementary formula of *Brevibacterium flavum* biomass, A, B, D, Y, (AN-Y), T — stoichiometric coefficients.

The response of C-limited *Brevibacterium flavum* chemostat culture to addition of various carbon substrates, consumed by cells, is similar. The amount of oxygen consumed for the oxidation of the equal doses of different carbon substrates varies. The normalized value β is introduced to express the efficiency of oxidation of the same amount of carbon substrate:

$$\beta = m_{O_2}^{exp}/m_{O_2}^{theor},$$

where: $m_{O_2}^{exp}$ — the measured quantity of oxygen, consumed for the oxidation of 1 mole of C-substrate;

$m_{O_2}^{theor}$ — the quantity of oxygen required for a complete oxidation of the same unit of C-substrate.

The value of β does not depend on biomass concentration. To our opinion β reflects the efficiency of the carbon substrate utilization by microorganisms in a certain physiological environment. Four different carbon sources were tested in *Brevibacterium flavum* chemostat culture and corresponding β values are given in Table 1.

Table 1. The efficiency of different carbon substrate utilization of *Brevibacterium flavum* chemostat culture ($D = 0.1\ h^{-1}$)

Carbon source	m_s [10^{-3} mol]	m_{O_2} [10^{-3} mol]	m_{O_2}/m_s [mol/mol]	β
Sucrose	1.389	5.080	3.660	0.305
Acetate	3.050	1.715	0.562	0.281
Succinate	4.237	3.640	0.859	0.245
Malate	3.731	8.432	2.260	0.753

The growth efficiency of *Brevibacterium flavum* on sucrose was studied in chemostat culture at different dilution rates (Table 2). β values were compared with the integral growth yield $Y_{x/s}$ and growth efficiency η. η was calculated according to EROSHIN and MINKEVICH [1]:

$$\eta = \frac{\sigma_b \gamma_b}{\sigma_s \gamma_s} Y_{x/s}^{exp}.$$

Table 2. The yield of biomass on sucrose $Y_{x/s}^{exp}$, thermodynamical efficiency of growth η, the coefficient β for *Brevibacterium flavum* chemostat culture

D [h^{-1}] → Coefficients ↓	0.154	0.220	0.302	0.380
$Y_{x/s}^{exp}$	0.31	0.39	0.44	0.49
η	0.42	0.53	0.60	0.67
β	0.273	0.256	0.213	0.204

There is an increase of $Y_{x/s}$ and η with the growth rate and decrease of β (Table 2). The efficiency of utilization of the auxiliary substrate — acetate also increases with the acceleration of medium flow rate (Fig. 3).

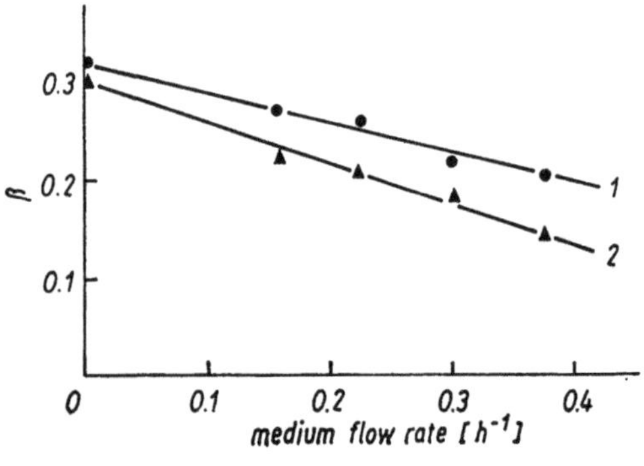

Fig. 3. The dependence of β on medium flow rate in *Brevibacterium flavum* chemostat culture
1 — sucrose (growth substrate),
2 — acetate (auxiliary substrate)

Discussion

The culture responds to the addition of the growth-limiting substrate with an increase of its metabolic activity. Such temporary activation can be followed by *pH* or pO_2 changes [4, 5], increase of respiration rate and heat generation [6, 7]. For the quantitative description of the culture response after the addition of a small carbon substrate dose the best results were obtained with the oxygen balance method [6]. Q_{CO2} measurement is considerably influenced by medium *pH* changes. Heat evolution kinetics is influenced by the inertia of our measurement system [6].

The relationship between consumed oxygen and added carbon substrate can be expressed by a straight line. A similar linear relationship between oxygen uptake and added reducing sugar amount was obtained in fed-batch culture of *Corynebacterium* sp. [8]. The linear relationship between two variables m_{O2} and m_s indicate that the stoichiometric coeffient B in the growth equation (1) does not depend on substrate dose. From the linear regression analysis of the data in Fig. 2 (with a 95% confidence limit) the respective values of B and β are: for sucrose B $= 3.64 \pm 0.03$ mol O_2/mol sucrose, $\beta = 0.30$; for acetate B $= 0.56 \pm 0.02$ mol O_2/mol acetate, $\beta = 0.28$.

The normalized coefficient β is equal to the ratio BOD/COD. A simple relationship can be derived between the value of β and thermodynamic efficiency of microbial growth η [9].

$$\beta = 1 - \eta \quad (3$$

Thus, the value $\beta =$ BOD/COD expresses the nonproductive energy consumption. β and η can theoretically assume any value between 0 and 1. When $\beta \rightarrow 0$, $\eta \rightarrow 1$, *i.e.*, all the chemical energy of carbon substrate is converted into biomass. The other extreme, when $\beta \rightarrow 1$, $\eta \rightarrow 0$. In this situation all the carbon substrate is completely oxidized into $CO_2 + H_2O$ and no chemical energy stored in the organic substrate is incorporated into biomass.

Experimentally the relation of β with the growth yield $Y_{x/s}$ and the calculated thermodynamical growth efficiency η is proved by the data in Table 2. In *Brevibacterium flavum* chemostat culture the increase of the medium flow rate is simultaneous with the increase of $Y_{x/s}$ and decrease of β. At any flow rate the experimental values of β are smaller than the calculated non-productive energy consumption $(1 - \eta)$. One of the reasons for such a discrepancy might be the extracellular product synthesis, especially at slower growth rates.

For microbial growth on various C-substrates the values of $Y_{x/s}$ are calculated using macroscopic principles [1, 2, 3, 10]. Theoretically the maximum yield values are approximated from the balance of redoxons [2, 3], mass and energy [1] or enthalpy [10]. From these principles it follows that the energy content per C-eq for sucrose and acetate is almost equal. Both substrates are of the same degree of reduction $\gamma = 4.0$ [2, 3], and with similar specific free enthalpy [2, 10]. However, the experimental values of β for sucrose and acetate are different (Fig. 2, 3, Table 1). One can conclude, that acetate ($\beta = 0.28$) is a more effective growth substrate in *Brevibacterium flavum* than sucrose ($\beta = 0.30$).

Conclusions

There are at least three advantages of the method described above:

A. Such fundamental parameter as the efficiency of microbial growth can be evaluated promptly. The growth efficiency of different carbon sources can be tested during a short period of time.

B. The value β reflects the virtual efficiency of carbon substrate utilization when the influence of both, cell metabolism and environment are taken into account.
C. A number of tentative technological parametres of the process can be tested indirectly, for example such as substrate consumption rate, heat generation efficiency, a.o.

Nomenclature

X — concentration of biomass, $g \cdot l^{-1}$
S — concentration of substrate, $g \cdot l^{-1}$
$Y_{x/s}$ — biomass yield on substrate, $g \cdot g^{-1}$
BOD — biochemical oxygen demand
COD — chemical oxygen demand
D — medium flow rate, h^{-1}
μ — specific growth rate, h^{-1}
q_s — specific rate of substrate consumption, $g \cdot g^{-1} \cdot h^{-1}$
β — normalized oxygen demand = BOD/COD, dimensionless
η — thermodynamic efficiency factor for microbial growth, dimensionless
γ — degree of reduction, eq redoxons/g · eq C
σ — carbon weight fraction, dimensionless
m_{O_2} — mass of oxygen, g mol
m_s — mass of substrate, g mol

Subscripts

x, b — biomass
s — substrate

Received April 15, 1985

References

[1] Eroshin, V. K., Minkevich, I. G.: Biotechnol. Bioeng. **24** (1982), 2263.
[2] Roels, J. A.: Energetics and Kinetics in Biotechnology. Amsterdam: Elsevier Biomedica 1983. 340 p.
[3] Minkevich, I. G.: Biotechnol. Bioeng. **25** (1983), 1267.
[4] Dudina, L. P., Eroshin, V. K.: Prikladn. Biokhimiya & Mikrobiologiya **15** (1979), 817 (in Russian).
[5] Dudina, L. P., Eroshin, V. K.: Mikrobiologiya **53** (1984), 791.
[6] Baburin, L., Shvinka, J. — In: Microbial Synthesis of Enzymes and Obtaining of its Preparative Forms. Riga: Zinatne, 1983. 78 (in Russian).
[7] Baburin, L. A., Shvinka, J. E., Viesturs, U. E.: Author's Certificate USSR, N 1150226 (1985).
[8] Ki Tae Hong, Sun Ho Park, Jae Heung Lee, Cha Yong Choi: J. Ferment, Technol. **62** (1984), 49.
[9] Esener, A. A., Roels, J. A., Kossen, N. W. F.: Biotechnol. Lett. **3** (1981), 193.
[10] Babel, W.: Z. Allg. Mikrobiol. **19** (1979), 671.

Acta Biotechnol. 6 (1986) 2, 129—132

Der substratlimitierte pH-Auxostat — Eine neue Methode zur kontinuierlichen Kultur

BÜTTNER, R., UEBEL, B., GENZ, I.

Ernst-Moritz-Arndt-Universität Greifswald
Sektion Biologie, WB Technische Mikrobiologie
DDR-2200 Greifswald, Jahnstraße 15a

Summary

A method for continuous cultivation of microorganisms is demonstrated, the substrate limited pH-auxostat. The limiting substrate only is added with constant velocity. In this culture the cells grow with high utilization of the limiting substrate and with the highest specific growth rate possible at the given conditions. Yield coefficients and dilution rates of stable K^+-limited steady states in yeast cultures with different pH-values and biomass concentrations were measured.

Zur Selektion und Untersuchung von Mikroorganismen für die Biomasseproduktion sind kontinuierliche Kulturen vom Turbidostat-Typen geeignet, da hier durch den Selektionsdruck diejenigen Zellen bevorzugt sind, die bei den vorgegebenen Bedingungen mit der höchsten spezifischen Wachstumsrate μ wachsen. Stabile Fließgleichgewichtszustände sind in kontinuierlichen Kulturen von diesem Typ aber nur möglich, wenn alle Komponenten des zufließenden Nährmediums im Überschuß vorliegen. Das gilt auch für die pH-Auxostat-Kulturen [1, 2], bei denen die Säureausscheidungsgeschwindigkeit als Maß für die Biomassebildungsgeschwindigkeit dient und zur Regelung des Mediumzuflusses entsprechend der aktuellen spezifischen Wachstumsrate der Kultur genutzt wird. In pH-Auxostat-Kulturen von Hefen fanden wir, abweichend von der in [2] aufgestellten Annahme, daß wachstumslimitierende Konzentrationen einer oder mehrerer Komponenten des Nährmediums stets zum Wachstumsstillstand der Kultur führten. Es sind daher zumindest für Hefen im pH-Auxostaten nur O_2-limitierte oder unlimitierte stabile steady states erreichbar. Da Substratüberschuß im industriellen Prozeß meist unerwünscht ist, wird die pH-Auxostat-Methode relativ selten zur Modellierung technischer Verfahren im Labor angewendet, obwohl sie einige Vorteile gegenüber dem Chemostaten aufweist. Das sind insbesondere die Selektion auf die unter den vorgegebenen Kultivierungsbedingungen maximal möglichen spezifischen Wachstumsraten μ_{max} und ihre direkte Messung, Vorauswahl der gewünschten Biomassekonzentration $\bar{x}$, keine Auswaschung der Biomasse bei Überschreitung der Stabilitätsgrenzen des Fließgleichgewichtszustandes. Wir entwickelten daher ein Kultivierungsverfahren, bei dem die Vorteile des pH-Auxostaten kombiniert sind mit dem Vorteil des Chemostaten, stabile steady states in substratlimitierten Kulturen untersuchen zu können.

Die Methode der substratlimitierten pH-Auxostat-Kultur

Die Besonderheit dieser Form der kontinuierlichen Kultur besteht darin, daß zwei Nährlösungen getrennt voneinander zugegeben werden. Die Lösung 1 enthält nur limitierendes Substrat und fließt (wie beim Chemostaten) mit konstanter Geschwindigkeit zu. Die Lösung 2 enthält (wie beim Chemostaten) alle übrigen essentiellen Substrate im Überschuß. Der Zufluß der Lösung 2 wird nach dem pH-Auxostat-Verfahren geregelt, indem bei der pH-Statierung der Kultur parallel zur Lauge Nährlösung zufließt. Dabei muß das Mengenverhältnis von Lauge zu Nährlösung konstant sein [3]. Eine Zugabe von Nährlösung und Lauge wird ausgelöst, nachdem im Fermentor so viel Biomasse neugebildet wurde, daß die dabei frei werdenden H^+-Ionen die vorher zugegebene Menge des Gemisches von Nährlösung und Lauge auf den Statier-pH-Wert angesäuert haben. Als Pufferungsfähigkeit BC [mMol H^+/l] wird die Menge H^+-Ionen definiert, durch die ein Liter des Gemisches auf den Statier-pH-Wert gebracht wird [2]. Im steady state müssen bei der Bildung von $\bar{x}$ g TG pro Liter BC mMol/l H^+-Ionen freigesetzt werden. Mit der Wahl des Mengenverhältnisses von Lauge zu Nährlösung wird die Biomassekonzentration

$$\bar{x} = Y_{H^+} \cdot BC \tag{1}$$

vorgegeben ($1/Y_{H^+}$ bezeichnet die Menge H^+-Ionen, die beim Statier-pH-Wert durch die Bildung von einem g Biomasse freigesetzt wird [2, 3]).

In der substratlimitierten pH-Auxostat-Kultur ist die Zuflußgeschwindigkeit F gleich der Summe der Zuflüsse F_1 (Lösung 1) und F_2 (Lösung 2 und Lauge), so daß die Durchflußrate $D = (F_1 + F_2)/V$ ist. Wir bezeichnen mit c_1 und c_2 die Konzentrationen des limitierenden Substrates, mit BC_1 und BC_2 die Pufferungsfähigkeiten in Lösung 1 und Lösung 2, die Größen F_1/V und F_2/V mit D_1 und D_2. Dann ist die Konzentration im gesamten Zufluß

$$S_{RL} = (D_1 \cdot c_1 + D_2 \cdot c_2)/D \tag{2}$$

und BC im Gesamtzufluß

$$BC = (D_1 \cdot BC_1 + D_2 \cdot BC_2)/D. \tag{3}$$

Die Substratbilanz für das limitierende Substrat L ist im steady state

$$S_{RL} - \bar{S}_L = \bar{x}/Y_L. \tag{4}$$

Unter Berücksichtigung von (1) nimmt sie die Form

$$D_1 \cdot c_1 + D_2 \cdot c_2 - \bar{S}_L \cdot D = (D_1 \cdot BC_1 + D_2 \cdot BC_2) \cdot Y_{H^+}/Y_L \tag{5}$$

an. In der Praxis sind meist die Bedingungen $c_2 = 0$ und $BC_1 = 0$ (pH-Wert der Lösung = Statier-pH-Wert) realisierbar, so daß die vereinfachte Gleichung

$$D_1 \cdot c_1 - \bar{S}_L \cdot D = BC \cdot Y_{H^+} \cdot D/Y_L \tag{6}$$

benutzt werden kann, da mit $BC_1 = 0$ $D_2 \cdot BC_2 = D \cdot BC$ gilt. Mit Gleichung (1) folgt daraus

$$D \cdot \bar{x} = Y_L(D_1 \cdot c_1 - \bar{S}_L \cdot D). \tag{7}$$

Für etwa konstante Y-Werte und $\bar{S}_L \ll S_{RL}$ muß dann $D \cdot \bar{x}$ etwa gleich $Y_L \cdot D_1 \cdot c_1$ sein, das heißt die Produktivität $D \cdot \bar{x}$ muß bei verschiedenen, durch BC vorgegebene Biomassekonzentrationen etwa gleich groß sein. Die Kurve $D = f(\bar{x})$ sollte entsprechend der Beziehung

$$D \approx Y_L \cdot D_1 \cdot c_1 \cdot 1/\bar{x} \tag{8}$$

etwa hyperbelförmig verlaufen.

Kaliumlimitierte pH-Auxostat-Kultur einer Hefe

Zur Erprobung der Methode wählten wir Kalium als limitierendes Substrat. Damit sollte gleichzeitig geprüft werden, ob auch bei der Einstellung der Zellen auf höchstmögliche Teilungsgeschwindigkeit die in [4] für Chemostat-Kulturen beschriebene Beziehung zwischen der spezifischen Wachstumsrate μ und dem Kaliumgehalt der Zellen auftritt.

Die Untersuchungen wurden am Hefestamm 217 (Stammsammlung, WB Technische Mikrobiologie) durchgeführt. Die Lösung 2 enthielt pro Liter 25 g $NH_4H_2PO_4$, 5 g $MgSO_4 \cdot 7\,H_2O$, 0,1 g $Ca(NO_3)_2 \cdot 4\,H_2O$, 0,01 g $FeCl_3 \cdot 6\,H_2O$, 2,5 ml der 100fach konzentrierten Spurenelementelösung nach [5] und 20 g Glukose. Die Lösung 1 enthielt 0,58 g KCl pro Liter. Die Kultivierung wurde in einem Eigenbaufermentor mit Rührung (Diplomarbeiten BRINKMANN, U., 1980; BÜTTNER, R., 1981; MUJIKA, G., 1982; alle Sektion Biologie der Ernst-Moritz-Arndt-Universität Greifswald) bei einem Kulturvolumen von 0,2 l, einer Belüftungsrate von 10–20 l/h, 32,5 °C, und bei pH-Werten zwischen 2,8 und 4,2 durchgeführt. Das Trockengewicht wurde nach Filtration (Synpor 6 Filter, Chemapol Praha) und die K^+-Konzentration flammenphotometrisch bestimmt. Batch-Versuche in diesem Fermentor mit dem Stamm 217 und Glukose als C-Quelle ergaben bei $\mu \cdot x$-Werten bis zu 1,5 g TG/l · h streng exponentielles, das heißt unlimitiertes Wachstum.

Wir erhielten unter allen geprüften Kulturbedingungen (Abb. 1) stabile steady states, die für jeweils sechs bis zehn Verdopplungszeiten aufrechterhalten wurden. Die Konzentrationen S_{RL} für Kalium betrugen 8 bis 48 mg K^+/l zufließender Gesamtlösung. Die nach (8) zu erwartende Hyperbelform für $D = f(\bar{x})$ ist in der Abb. 1 A angedeutet. Dieser vom theoretischen Verhalten unlimitierter pH-Auxostat-Kulturen ($D = \mu_{max}$, unabhängig von $\bar{x}$ [2]) abweichende Kurvenverlauf ist jedoch auch bei O_2-Limitation und bei Produkthemmung zu erwarten, das heißt bei hohen $\bar{x}$ beziehungsweise $D \cdot \bar{x}$ Werten. Die $D \cdot \bar{x}$-Werte der hier untersuchten steady states liegen zwischen 0,3 und

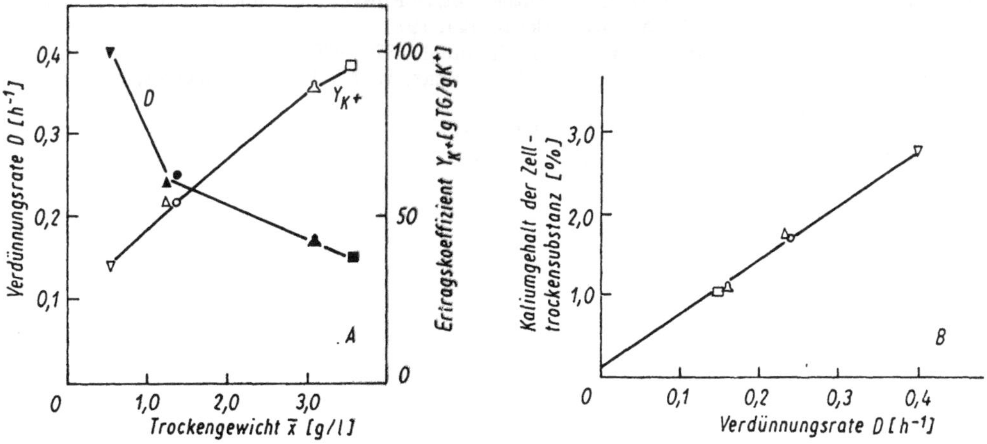

Abb. 1. Beziehungen zwischen Durchflußrate D, Biomassekonzentration $\bar{x}$ und dem Kaliumgehalt der Biomasse in einer Kalium-limitierten pH-Auxostat-Kultur einer Hefe. In der Abbildung sind die Mittelwerte aus den in jeweils einem steady state erhaltenen Meßdaten eingezeichnet. Es wurden steady states mit einer Pufferungsfähigkeit des Mediums (BC) von 20 mMol/l bei pH 2,8 (▽—), pH 3,0 (△—), pH 3,7 (○—) und mit 30 mMol/l bei pH 3,7 (○—) und pH 4,2 (□—) eingestellt.

1 A: Abhängigkeit der Durchflußrate und des Ertragskoeffizienten Y_{K^+} von der Biomassekonzentration,

1 B: Kaliumgehalt der Trockensubstanz in Abhängigkeit von D (nach NYHOLM)

0,7 g TG/l · h. Die Abnahme von D mit steigendem $\bar{x}$ dürfte daher zumindest vorwiegend auf die K^+-Limitation zurückzuführen sein, auch wenn der O_2-Bedarf der K^+-limitierten Zellen erhöht ist. Dafür sprechen auch die hohen, mit sinkendem D zunehmenden Werte für Y_{K^+} (Abb. 1A). Deutlicher kommt dieser Zusammenhang in der Abb. 1B zum Ausdruck. Die nach [4] aufgetragenen Werte ergeben die Gerade $10/Y_{K^+} = 6{,}7\,D + 0{,}1 = \%$ Kalium in der Trockensubstanz. Zur Berechnung der Regressionsgeraden wurden die in den steady states erhaltenen Einzelwerte benutzt, wobei sich ein Korrelationskoeffizient von 0,97 bei einer Konfidenzgrenze $> 99{,}9$ ergab. Danach beträgt der Mindestgehalt an Kalium in nicht wachsenden Zellen 0,10 g K^+ pro 100 g TG. Der streng lineare Zusammenhang zwischen Kaliumgehalt und der spezifischen Wachstumsrate der Zellen entspricht der in [4] aufgestellten und an Chemostat-Kulturen überprüften Hypothese, daß die spezifische Wachstumsrate bei Limitation durch konservative Substrate, wie zum Beispiel K^+, vorwiegend durch den Gehalt in den Zellen und nicht durch die Außenkonzentration $\bar{S}_L$ bestimmt ist.
Aus der Gleichung (7) ist weiterhin ersichtlich, daß für gleiches $\bar{x}$ der K^+-Gehalt in den Zellen und damit D ansteigt, wenn $\bar{S}_{K^+}$ kleiner wird. Im mediumlimitierten pH-Auxostaten werden also Zellen selektioniert, die bei dem höchsten Y_L-Wert und dem kleinsten $\bar{S}_L$-Wert mit der höchsten spezifischen Wachstumsrate wachsen. Das ist ein wesentlicher Unterschied zum Chemostaten, in dem diejenigen Zellen ausgelesen werden, die bei der vorgegebenen, deutlich unter μ_{max} liegenden Wachstumsrate bei dem kleinsten $\bar{S}_L$-Wert wachsen können, wobei der Y_L-Wert keinen Einfluß hat [1].

Eingegangen: 22. 8. 1984

Literatur

[1] Petschurkin, N. S., Terskow, I. A.: Autoselektive Prozesse in kontinuierlicher Kultur von Mikroorganismen (russisch). Nowosibirsk: Nauka, 1973.
[2] Martin, G. A., Hempfling, W. P.: Arch. Microbiol. **107** (1976), 41.
[3] Oltman, L. F., Schoenmaker, G. S., Reijnders, W. N. M., Stouthamer, A. H.: Biotechnol. Bioeng. **20** (1978), 921.
[4] Nyholm, N.: Biotechnol. Bioeng. **18** (1976), 1043.
[5] Fritsche, W.: Z. Allg. Mikrobiol. **8** (1968), 91.

Acta Biotechnol. 6 (1986) 2, 133—139

Ein neues System von Begasungsbrutschränken für die Biotechnologie

BECKMANN, D., METZE, J.

Akademie der Wissenschaften der DDR
Zentrum für wissenschaftlichen Gerätebau, mytron
DDR-5630 Heilbad Heiligenstadt, Rosenhof

Summary

The experiences in the production of CO_2-incubators [1] have been the basis for the development of a system of gassing incubation cabinets, which, starting from a basic structure with temperature and humidity control (climatic test cabinets), consist of modular built up devices with regulating circuits for carbon dioxide and oxygen content of the incubation room atmosphere. The performance of the system is considerably extended by the equipment of the devices with a microcomputer system, by which the dynamic control of the parameters of the incubation room atmosphere is possible the first time.
Essential mark of the system is further the standardized structural module for all devices and the flexibility in the equipment of the incubation room according to the demands of special tests.

Jede Aktivität innerhalb der biotechnologischen Forschung zielt auf die effektive industrielle Nutzung lebender Mikroorganismen, pflanzlicher oder tierischer Zellen bzw. ihrer Produkte. Dabei steht am Anfang der Einbindung biologischer Prozesse in großtechnische Verfahren stets die Bereitstellung geeigneter Mikroorganismen bzw. Zellstämme.

Die Umsetzung neuer wissenschaftlicher Erkenntnisse und Methoden schließt das Vordringen zu einzelnen Zellen und zum Zellinneren ein. Damit kommt es zur Zerstörung der natürlichen Schutz- und zur Einschränkung der Lebensmechanismen der Zellen. Kultivierung von Zellen im Rahmen mikrobiologischer Forschung setzt deshalb gerätetechnische Apparaturen voraus, die in vitro eine aseptische Umgebung und optimale Lebensbedingungen für das empfindliche Zellmaterial garantieren. Neue Kultivierungsmethoden müssen daher auf höherem technischen Niveau realisiert werden. Dieser Anspruch wird unterstrichen durch die hohen Kosten, die im Rahmen der Kultivierung bzw. beim Kauf von Leistungsstämmen anfallen. Die Basisapparatur zur Kultivierung von Zellen in diesem Stadium ist der Begasungsbrutschrank.

Ausgehend von den gesammelten Erfahrungen bei der Produktion und Nutzung von CO_2-Inkubatoren [1] wurde entsprechend den Fortschritten auf dem Gebiet des wissenschaftlichen Gerätebaus ein neues System von Begasungsbrutschränken sbs entwickelt. Aufbauend auf einer einheitlichen Gerätegrundstruktur, die mit einem durchgängigen Rastermaß universell einsetzbar ist und entsprechend den Anforderungen der Nutzer eine modulare Ausrüstung bzw. eine komplikationslose Nachrüstung ermöglicht, beinhaltet das Gerätesystem vier Grundvarianten mit einem Nutzraumvolumen von 115 l, die sich wiederum in spezifische Ausführungsformen unterteilen lassen (Abb. 1).

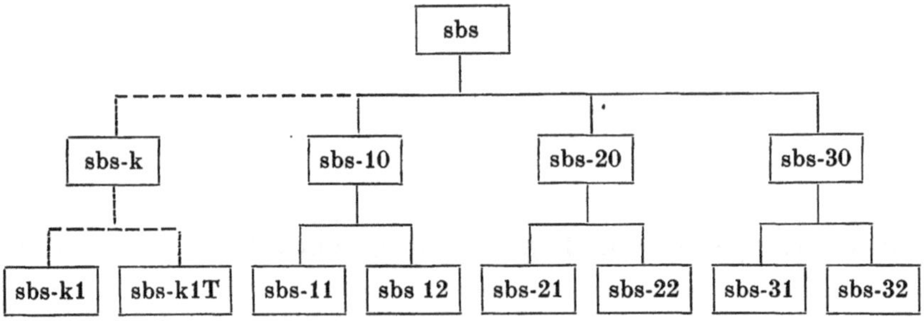

	sbs-k1	sbs-k1T	sbs-11	sbs 12	sbs-21	sbs-22	sbs-31	sbs-32
Geräteart	Klimaprüfschränke		Begasungsbrutschränke mit analogen Meß-, Steuer- und Regelsystemen		Begasungsbrutschränke mit Rechnersteuerung		Gen-Arbeitskomplexe mit Analogelektronik	mit Rechnersteuerung
							Kombination von Begasungsbrutschrank und Laminarsicherheitsbox	
Ausrüstung								
Temperaturregelung	•	•	•	•	•	•	•	•
Feuchteregelung	•	•	•	•	•	•	•	•
CO_2-Regelung	—	—	•	•	•	•	∘	∘
O_2-Regelung	—	—	—	•	—	•	—	∘
Ansteuerung von Optionsbaugruppen	—	—	∘	∘	•	•	∘	∘

• vorhanden ∘ möglich — nicht vorhanden

Abb. 1. Das Gerätesystem sbs

1. Klimaprüfschränke
   Klimaprüfschränke wurden wegen der einheitlichen Grundstruktur in das System sbs einbezogen
2. Begasungsbrutschränke mit analogen Regelsystemen (sbs-10)
3. Begasungsbrutschränke mit Geräterechner (sbs-20)
4. Gen-Arbeitskomplex sbs-30
   Kombination eines Begasungsbrutschrankes sbs mit einer Laminarsicherheitsbox

Im folgenden soll nur auf die Typen sbs-10 und sbs-20 näher eingegangen werden. Abb. 2 zeigt schematisch die perspektivische Ansicht einer Gerätevariante.
Alle Geräte des Systems besitzen standardmäßig elektronische Regelsysteme für die Parameter Temperatur und relative Luftfeuchte. Die technischen Daten der Geräte sind in Tabelle 1 angegeben. Die Varianten sbs-10 und sbs-20 sind je nach Ausführungsform mit einem Regelsystem für die Größen Kohlendioxid-Gehalt der Brutraumatmosphäre (sbs-11, sbs-21) und Sauerstoff-Gehalt der Brutraumatmosphäre (sbs-12, sbs-22) ausgerüstet und können mit einem System zur Ansteuerung von Optionsbaugruppen (Frequenz bzw. Drehzahl von Magnetrührern oder Flaschenrollern) erweitert werden.

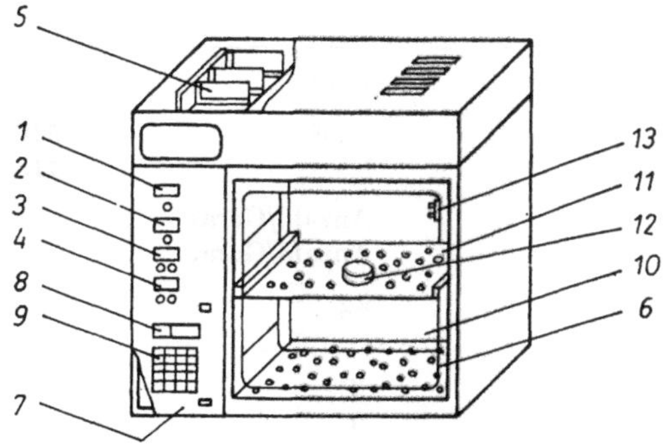

Abb. 2. Perspektivische Ansicht einer Gerätevariante des Systems sbs

1 Temperaturanzeige, 2 Anzeige der relativen Luftfeuchte, 3 Anzeige der CO_2-Atmosphäre, 4 Anzeige der O_2-Atmosphäre, 5 Geräterechner, 6 universelles Haltesystem, 7 Frontblende, 8 Anzeige für Optionsbaugruppen und Rechner, 9 Bedientastatur, 10 Brutraum, 11 Einlegerost, 12 biologisches Untersuchungsobjekt, 13 Sensoren

Die Anzeige der Soll- und Istwerte erfolgt generell digital und für jeden Parameter separat. Die Bedienelemente (Potentiometer bzw. Tastatur) befinden sich übersichtlich auf einem Bedientableau auf der linken Seite der Vorderfront.

Der Brutraum ist mit einer verriegelbaren Gasblende abgeschlossen und verhindert einen schnellen Gasaustausch mit der Umgebung nach Öffnen der äußeren Gerätetür, ermöglicht jedoch durch seine Ausführung als Glastür den Einblick in den Brutraum ohne Beeinflussung der Untersuchungen.

Alle Innenflächen des Brutraumes, das Haltegestell und die Einlegeböden bestehen aus rost- und säurebeständigem Stahl, sind durch ihre Konstruktion einfach zu reinigen und gewährleisten eine leichte Desinfektion und sterile Betriebsweise.

Für das Gerätesystem sbs wurden ein flexibles Haltegestell entwickelt [2], das die Einlegeböden kippsicher aufnimmt und sowohl die Möglichkeit einer universellen Lagefixierung dieser Böden als auch die komplikationslose Unterbringung der o. g. Optionsbaugruppen bietet. Nachfolgend sollen die einzelnen Regelsysteme vorgestellt werden.

Temperaturregelung

Über das entsprechende Bedienelement (Potentiometer oder Tastatur) ist es möglich, den gewünschten Sollwert und den Übertemperatursollwert einzustellen. Bei Erreichen des Übertemperatursollwertes werden zum Schutz des Arbeitsgutes vor Überhitzung, z. B. bei einer Havarie, alle Wärmequellen blockiert.

Die Temperierung des Brutraumes erfolgt sowohl im Innenraum als auch in der Tür des Gerätes über Großflächenheizkörper.

Zusammen mit einer turbulenzarmen vertikalen Luftführung (max. 0,5 — 0,6 m/s) wird eine gleichmäßige räumliche und zeitliche Temperaturverteilung erreicht und die Kondensatbildung an der Decke und den Wänden des Brutraumes verhindert.

Hinter der Rückwand des Brutraumes wurde eine Kühl- und Solltaufläche installiert, die jedoch nur bei Überschreitung des Übertemperatursollwertes in Betrieb genommen wird. Die Regelcharakteristik des Temperaturregelkreises ist so optimiert, daß eine Kühlung während eines Betriebsregimes mit Konstanttemperatur nicht erforderlich ist. Wegen des physikalischen Zusammenhangs von Temperatur und Feuchte werden durch diese Maßnahme negative Auswirkungen auf die Feuchteregelcharakteristik vermieden.

Tabelle 1. Technische Daten der Geräte des Systems sbs

Außenmaße B × H × T	mm^3	900 × 920 × 600
Innenraum B × H × T	mm^3	540 × 540 × 400
Volumen	l	115
Einschubböden serienmäßig	Anzahl/Gerät	3
maximal	Anzahl/Gerät	5
Masse	kg	ca. 150
Temperaturregelung		
Bereich	°C	20[1])···70
mit Hermetikverdichter erweiterbar auf	°C	−5···70
räumliche Temperaturkonstanz	K	±0,4
zeitliche Temperaturkonstanz	K	±0,3
Befeuchtung		
Bereich	% r. F.	62···98
Toleranz	%	±3,5
CO_2-Regelung		
Bereich	% CO_2	0···20
Erholzeit	% CO_2/min	1
Fehler der gesamten Meßstrecke (bezogen auf den Meßwert)	%	≤ 4
O_2-Regelung		
Bereich	% O_2	0···21
Desinfektionsart	chemische Desinfektion	
Anschlußmöglichkeit für Optionsbaugruppen Optionsbaugruppen	Steckverbinder	
Hardwareausrüstung (sbs-20)		
CPU Prozessor U 880		
11 kbyte EPROM		
5 kbyte RAM (dav. 1 kbyte CMOS)		
5 PIO		
1 CTC		
Schutzeinrichtungen		
Temperaturüberschreitung	Meßsystem mit Fühlerausfallerkennung und optischer Meldung	
Reduzierung des Wasservorrates für Feuchteregelung auf 1 l	optische Meldung	
Wasservorrat erschöpft	optische Meldung	
Über- und Unterschreitung des CO_2-Sollwertes	optische Meldung	
Über- und Unterschreitung des O_2-Sollwertes	optische Meldung	
Geräteschutz	Temperaturwächter, einstellbar	
Ausgabe einer Sammelstörmeldung für externen Anschluß		
Ausgabe aller Istwertpegel für externen Geräteanschluß	0···10 V, Schnittstelle V 24 bei Rechnervarianten	
Schutzgrad	IP 20	
Schutzklasse	I	

[1]) unterer Temperatursollwert mindestens 10 K über der Raumtemperatur.

Feuchteregelung

Alle Geräte des Systems besitzen eine Einrichtung zur Intensivbefeuchtung der Schrankatmosphäre; dadurch wird die Austrocknung von Kulturen in offenen Gefäßen (z. B. Petrischalen) verhindert. Der Feuchtesollwert wird mit dem entsprechenden Bedienelement als Führungsgröße für den Regelkreis eingestellt. Die Feuchtemessung erfolgt mit einem elektrolytischen Leitfähigkeitssensor. Generell hat die Regelung der relativen Feuchte Priorität vor der Regelung der Temperatur.
Damit wird ein Überschwingen der Temperatur infolge des Wärmeeintrags durch den Verdampfer für destilliertes Wasser vermieden.

Regelung des CO_2-Gehaltes

Der CO_2-Gehalt der Nutzraumatmosphäre wird mit dem Infrarot-Gasanalysator Infralyt [3] gemessen. Die Zuführung von Kohlendioxid erfolgt über ein Magnetventil, das taktweise betätigt wird, aus einer Flüssiggasflasche.

Regelung des Sauerstoffgehaltes

Zur Messung des Sauerstoffgehalts wird ein elektrochemischer Sauerstoffsensor (CLARK-Prinzip, Forschungsinstitut Meinsberg [4, 5]) eingesetzt. Der Regler entspricht weitestgehend dem CO_2-Regler. Die Regelung des am Bedienelement eingestellten Sollwertes erfolgt durch Zuführung von Luft oder Stickstoff in Abhängigkeit vom Vorzeichen der Regelabweichung. Beide Gase werden über Magnetventile aus Flüssiggasflaschen entnommen.

Desinfektion

Die Begasungsbrutschränke des Systems sbs besitzen alle notwendigen Voraussetzungen für die chemische Desinfektion [6]. Aufgrund der guten Ergebnisse und der Vorteile dieser Methode, die schon beschrieben wurde [1], wird dieses Verfahren auch für das System sbs eingesetzt. Das Desinfektionsmittel wird in einer offenen Schale in den Innenraum des Inkubators gestellt und verdampft dort. Nach der Desinfektion kann das Gas-Luft-Gemisch mit einer Taumelscheibenpumpe abgesaugt werden. Untersuchungen ergaben, daß die in Tabelle 2 aufgeführten Desinfektionsmittel eine sichere Keimabtötung ermöglichen.

Tabelle 2. Beispiele für verwendbare Agenzien zur Desinfektion der Begasungsbrutschränke sbs

Medium	Konzentration
Formalinlösung	$> 0{,}5\%$
Wofaseptlösung	$> 2{,}0\%$
Perchloressigsäure	$> 5{,}0\%$

Rechnersteuerung (sbs-20)

Entsprechend den hohen Anforderungen der Grundlagen- und angewandten Forschung an moderne Begasungsbrutschränke wurden die Geräte des Typs sbs-20 mit einem Gerätерechner zur Steuerung aller bereits genannten Parameter und zur automatischen Desinfektion des Brutraumes ausgerüstet. Die Struktur des Regelsystems und die Stellung des Mikrorechners ist in Bild 3 dargestellt. Der Rechner ermöglicht neben dem normalen Arbeitsregime, bei dem jederzeit über die Tastatur in den Prozeß eingegriffen

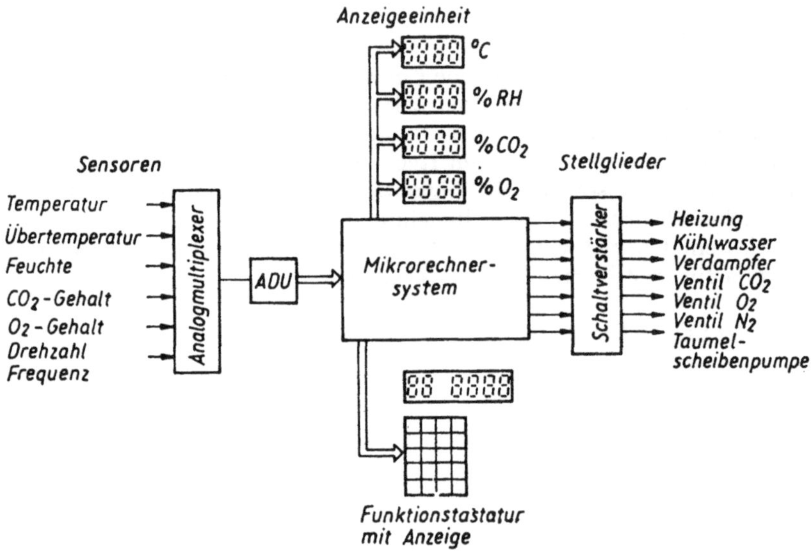

Abb. 3. Regelsystem der rechnergesteuerten Begasungsbrutschränke des Grundtyps sbs-20.

werden kann, die zeitabhängige Vorprogrammierung der regelbaren Parameter. Dadurch können beliebige dynamische Versuchsabläufe bis hin zur Desinfektion automatisch und ohne manuellen Eingriff abgefahren werden.

Die aufbereiteten und normierten Meßwerte (0—10 V) werden über den Analogmultiplexer und den ADU (10 bit) dem Mikrorechner zugeführt, der alle Steuer-, Regelungs- und Anzeigefunktionen gemäß der implementierten Software übernimmt. Die Kommunikation mit dem Bediener wird über eine Funktionstastatur und zugehörige 7-Segment-Anzeige (sechsstellig) realisiert. Mit dieser Bedieneinheit ist es jederzeit möglich, alle Prozeßdaten (Sollwerte, Istwerte, Prozeßzeiten) abzurufen bzw. zu beeinflussen. Die aktuellen Meßwerte von Temperatur, Feuchte, CO_2- und O_2-Gehalt werden ständig auf den jeweiligen Anzeigeeinheiten dargestellt. Der Mikrorechner stellt TTL-Signale zur Ansteuerung der Stellglieder zur Verfügung. Alle veränderlichen Prozeßdaten (Sollwerte, Prozeßzeiten, Arbeitsregime) werden in einem batteriegepufferten CMOS-RAM abgelegt und damit gegen Verlust bei kurzzeitigen Netzausfällen gesichert, womit ein Wiederanlauf mit den vorgegebenen Parametern gewährleistet wird.

Entsprechend den vorgegebenen Aufgaben ist die eingesetzte Software modular und strukturiert aufgebaut und damit weitestgehend voneinander unabhängig und universell verwendbar. Sie ist schwerpunktmäßig wie folgt gegliedert:

Systemkern
- Initialisierung (Systemstapel, Peripheriebausteine)
- Aufbau der Basisdatentabellen
- Start Echtzeituhr
- Start Echtzeitmonitor

periodisch aktivierte Programme mit fester Periodenzeit und Prioritätenfolge
- Meßwerterfassung (Meßstellenabfrage, Signalaufbereitung, Sinnfälligkeitskontrolle, Berechnung der physikalischen Größen, Normierung)
- digitale Regelung der Parameter Temperatur, relative Feuchte, CO_2-Atmosphäre, O_2-Atmosphäre, Drehzahl bzw. Frequenz der Optionsbaugruppen

über die Funktionstastatur anwählbare Programme (Hintergrundprogramme)

— Abfrage/Eingabe/Änderung von Sollwerten
— Definition und Auslösung von Zeitprofilen
— Eichen der CO_2- und O_2-Meßverstärker
— Ein-/Ausschalten der Meßwerterfassung und der Regelprogramme
— Auslösung und Überwachung der automatischen Desinfektion

Gerätetreiberprogramme

— Zeicheneingabe Funktionstastatur
— Zeichenausgabe Anzeigezeile Funktionstastatur
— Zeichenausgabe Anzeigeeinheiten

Die vorgestellten Begasungsbrutschränke des Systems sbs ermöglichen eine sichere und zuverlässige Züchtung von Mikroorganismen und Zellkulturen. Mit diesen Geräten und den Erweiterungsmöglichkeiten innerhalb des Systems, insbesondere durch die Rechnersteuerung, werden den Wissenschaftlern der verschiedenen Disziplinen der Biologie moderne Forschungsgeräte für ein breites Aufgabenspektrum zur Verfügung gestellt [7, 8].

Eingegangen: 16. 1. 1985

Literatur

[1] BECKMANN, D., METZE, J., RHEINLÄNDER, H.: Acta Biotechnol. 4 (1984) 2, 187.
[2] METZE, J., HILDEBRANDT, E. u. a.: Haltegestell für Nutzräume von Umweltsimulationsgeräten, DD-WP C 12 M 2612518.
[3] Bedienungsanleitung Infralyt, VEB Junkalor Dessau.
[4] SPIEGEL, K.: msr 22 (1979) 5, 267.
[5] GÜNTHER, L., MÖHLER, W., SPIEGEL, K.: Elektrochemische Sauerstoffsensoren nach dem CLARKschen Prinzip der Prozeßüberwachung, Firmenschrift des Forschungsinstituts Meinsberg, 1979.
[6] JACOBI: Verfahren zur Kaltsterilisation mit einem Sterilisationsmitteldampf, DD-WP A 61 L/205160.
[7] METZE, J., KLEPPE, H.-G. u. a.: Verfahren und Vorrichtung zur Umweltsimulation für biologische Prozeßabläufe. DD-WP C 12 M 261250 1.
[8] METZE, J., BECKERT, R. u. a.: Gerät zur Umweltsimulation für biologische Prozeßabläufe, Musteranmeldung Akt.-Z. 15159-U 3545.

Acta Biotechnol. 6 (1986) 2, 140

Book Reviews

C. Buschmann, K. Grumbach

Physiologie der Photosynthese

Berlin, Heidelberg, New York, Tokyo: Springer-Verlag, 1985. 245 S., 106 Abb., 34 DM.

Im vorliegenden Buch geben die Autoren eine Übersicht über den derzeitigen Stand der Photosyntheseforschung, deren biophysikalische und biochemische Grundlagen sowie physiologische und ökologische Aspekte. Der Bogen der Themen der einzelnen Kapitel spannt sich von Struktur und Funktion photosynthetisch aktiver Zellen über Chemosynthese und anthropogene Faktoren der Photosynthese bis zu biotechnischen Ansätzen zur Nutzung der Photosynthese.

Die einzelnen Abschnitte werden durch Anleitung für zahlreiche Versuche ergänzt, die dem Leser eine sinnvolle Verknüpfung von Theorie und Praxis bieten und das Buch zu einem wertvollen Leitfaden sowohl für die Lehre als auch für einen Einstieg in die Forschung machen. Entsprechend dem Inhalt des jeweiligen Kapitels werden dabei Arbeitsvorschriften sowohl für inzwischen als „klassisch" anzusehende Versuche (z. B. Hill-Reaktion) als auch für Experimente vorgelegt, die mit neuen Entwicklungen verbunden sind (Messung der Wasserstoffproduktion bei Blaualgen).

Erfreulich für den Leser ist der klare Sprachstil, mit dem der Stoff behandelt wird.

Leider sind nicht alle graphischen Darstellungen gleich gut lesbar.

Am Ende der Kapitel sind Originalarbeiten und zusammenfassende Darstellungen (teilweise von 1984) gegeben, die den Interessenten ein vertieftes Studium der einzelnen Sachgebiete ermöglichen. Leider beziehen sich diese Arbeiten nicht auf die Literaturhinweise im Text, deren Wert dadurch weitgehend in Frage gestellt ist.

Den Schlußteil des Buches bildet eine Zusammenstellung von 50 wichtigen Fachbüchern zum Thema Photosynthese bzw. mit ausführlichem Photosynthese-Teil.

Insgesamt schließt das vorliegende Buch eine zweifellos vorhandene Lücke in der deutschsprachigen Darstellung des Gesamtgebietes Photosynthese. Die durch die Fülle bedingte gedrängte Behandlung des Stoffes macht es zu einem wertvollen Hilfsmittel für einen Einstieg in das Thema suchende Leser, ermöglicht aber auch bereits an speziellen Aspekten tätigen Wissenschaftlern eine Einordnung ihrer Arbeiten in das Gesamtgebiet der Photosynthese. Das Buch ist deshalb jeder sich mit Photosynthese im weitesten Sinne befassenden wissenschaftlichen Einrichtung zu empfehlen.

P. Roth

M. E. Beker u. a.

Transformazija produktov fotosintesa

(Transformation von Produkten der Fotosynthese)

Riga: Sinatne, 1984. 249 S.

Die von profilierten Wissenschaftlern des Mikrobiologischen Instituts der Lettischen Akademie der Wissenschaften herausgegebene Monographie informiert über den gegenwärtigen Stand der Forschungsarbeiten zur Biokonversion pflanzlicher Rohstoffe und die volkswirtschaftliche Umsetzung dieser Ergebnisse. Im einzelnen werden nicht-traditionelle Wege der Ethanolerzeugung, die Herstellung von polysaccharidspaltenden Enzymen (Amylasen, Cellulasen) und mikrobiellem Protein, die Biokonversion von Lignocellulosen (insbesondere von Stroh), die Gewinnung von Pflanzenprotein durch fermentative und thermische Fraktionierungsprozesse sowie die Erzeugung von Biogas behandelt. Ein abschließendes Kapitel ist der biologischen und zootechnischen Bewertung der Produkte der Biokonversion geschenkt.

Den untersuchten Mikroorganismen, ihrem Stoffwechsel und dessen Regulation sowie den isolierten Enzymen wird besondere Aufmerksamkeit gewidmet.

Aktuelle Literaturverzeichnisse sind jedem Kapitel zugeordnet und erlauben eine weitergehende Information. Eine Vielzahl graphischer Darstellungen, Tabellen sowie elektronenmikroskopischer Aufnahmen sind in den Text eingefügt.

Das Buch belegt eindrucksvoll die Position der Biotechnologie in der landwirtschaftlichen Produktion. Die Monographie ist nicht nur Mikrobiologen, Biochemikern und Biotechnologen zu empfehlen, sondern auch Agrar- und Veterinärwissenschaftlern, die an der Verwertung pflanzlicher Biomassen und der Nutzung der biotechnisch erzeugten Wertstoffe interessiert sind.

G. Klappach

Acta Biotechnol. 6 (1986) 2, 141–146

Biotransformations.
XIX. Reduction of Some Terpenic Ketones by means of Immobilized Cells of Rhodotorula mucilaginosa*

MIRONOWICZ, A., SIEWINSKI, A.

Institute of Fundamental Chemistry
Agricultural University
50-375 Wrocław, ul. Norwida 25, Poland

Summary

Reduction of (–)-menthone ((–)-**1**), (+)-(R)-methyl-α-campholenone ((+)-**2**), (+)-carvone ((+)-**3**), and eucarvone (**4**) was carried out by means of cells of the *Rhodotorula mucilaginosa* species immobilized in polyacrylamide gel. Alcohols with the (S)-configuration, (+)-neomenthol ((+)-**1a**), (+)-(R)-methyl-α-campholenol ((+)-**2a**), (–)-neoisodihydrocarveol ((–)-**3a**), dihydroeucarveol ((–)-**4a**), and small amounts of (–)-dihydroeucarvone ((–)-**5**), were obtained. The cells of *R. mucilaginosa* maintained after this reaction ability to reduce standard acetophenone to (–)-1-phenyl-1-ethanol.

Introduction

Microbiological reduction of cyclic ketones belongs to the best recognized and controlled processes brought into practice in organic synthesis [1, 2]. To carry out such transformations, continuous methods using immobilized cells of microorganisms are more and more widely applied [3, 4], besides the so-called submerged fermentation. Both methods were subject of our interest. We applied them for transformation of selected ketones, being reduced to (S)-alcohols by the use of the *Rhodotorula mucilaginosa* species [5–8]. Continuing our studies on capability of this species to reduce stereospecifically the carbonyl group, we carried out the submerged transformation of such substrates as (+)–(R)-methyl-α-campholenone [9], (–)-carvone, and eucarvone [10]. Simple picture of these reactions prompted us to their repetition by the continuous method using immobilized cells of the species mentioned. Preliminary results of these experiments were described earlier [11].

Results and Discussion

The studied strain from the *Rhodotorula mucilaginosa* species was grown on a maltose nutrient as cell suspension suitable for immobilization. It was immobilized after centrifugation by the MOSBACH method in polyacrylamide gel [3]. These cells amounted appro-

* Part XVIII: Phytochemistry (1985), in the press.

ximately 15% of the polymerized mass which was used for filling the connected in series segments of an arrangement applied for continuous transformation.
For basic investigations, substrates from the monoterpene series or their close homologues, (—)-menthone ((—)-**1**), (+)—(R)-methyl-α-campholenone ((+)-**2**), (+)-carvone ((+)-**3**), and eucarvone (**4**), were used. The transformation of the first substrate (—)-**1** was carried out parallel by the submerged method.
Substrates for transformation were prepared in solution of phosphate buffer in concentration of ca. 100 mg of compound per 1 l. Low concentrations were due, among others, to low solubility of substrates in water. The solution was filtered within definite time through columns packed with immobilized cells of *R. mucilaginosa*. The reaction was controlled by means of gas chromatography and its results were confirmed spectrally by investigation of products and unreacted substrates isolated by column chromatography. The transformation was carried out until gas chromatography showed three-four-fold decrease of the yield of expected product.

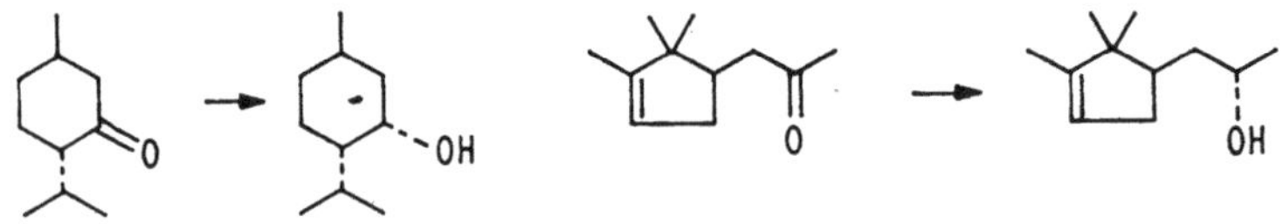

(-)-1 (+)-1a (+)-2 (+)-2a

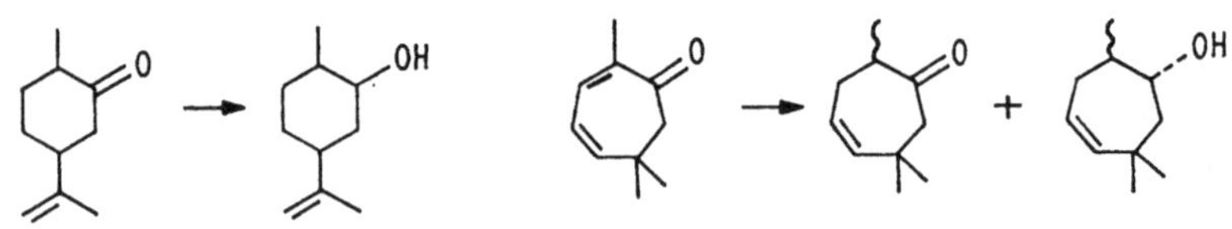

(+)-3 (+)-3a 4 (-)-5 (-)-4a

Fig. 1. Reduction of selected substrates by means of immobilized cells of *Rhodotorula mucilaginosa*

Fig. 1 presents the scheme of reactions carried out by the continuous method. Products formed were qualitatively consistent with products of the submerged transformation [9, 10]. A quantitative difference concerned eucarvone [4] which afforded (—)-dihydroeucarvone ((—)-**5**) by the submerged method whereas in the continuous method the reaction run further to (—)-dihydroeucarveol ((—)-**4a**) as the main product. All alcohols formed had the (S)-configuration. (+)-Neomenthol ((+)-**1a**) and (+)—(R)-methyl-α-campholenol ((+)-**2a**) were obtained from substrates (—)-**1** and (+)-**2**, respectively. In the case of transformation of substrates (+)-**3** and **4**, reduction of the double bond was observed preceding reduction of the carbonyl group: (—)-neoisocarveol ((—)-**3a**) and (—)-dihydroeucarveol ((—)-**4a**) were formed.
Tab. 1 shows conditions of the continuous transformation of substrates to be reduced and yields of products determined by means of GC. These yields were calculated as average values from all fractions obtained from column after transformation.
In the case of substrate (+)-**3**, after its transformation column was washed with buffer solution, stored at 278 K for 10 days and used again for transformation of acetophenone as the standard substrate. It underwent reduction to (—)-1-phenyl-1-ethanol within nine days with a 100% yield. The most important of our observations made hitherto seems to be this one that cells of the *Rhodotorula mucilaginosa* species in the immobilized

Table 1. Comparison of continuous transformations of four substrates by means of cells of *Rhodotorula mucilaginosa* immobilized in polyacrylamide gel

Substrate	Composition of isolated mixture of products	Conditions of transformation				
		Time	Volume of substrate solution	Total amount of substrate	Volume of centrifugated cells of *Rh.m.*	Flow time of substrate through column
		days	l	mg	ml	h
(−)-Menthone (−)-**1**	40% of (−)-**1** 40% of (+)-**1a** 20% — others	5	5	500	7	6
(+)-(R)-Methyl-α-campholenone (+)-**2**	70% of (+)-**2a** 20% of (+)-**2** 10% — others	2	1.8	180	14	12
(+)-Carvone (+)-**3**	30% of (+)-**3** 60% of (−)-**3a** 10% — others	14	8	1200	14	12
Eucarvone **4**	35% of **4** 55% of (−)-**4a** 5% of (−)-**5**	8	7.3	730	28	24

form do not lose their ability to transform selected ketones. Moreover, they retain their activity even during repeated application for another substrat. Furthermore the reaction time is relatively short.

Materials and Indentification of Products

1. *Biological Material*

Growth of the *Rhodotorula mucilaginosa** culture was carried out according to [12]. Preliminary culture was prepared in 100 cm³ of 2% maltose nutrient and it was shaken for one day at 300 K Then the culture was transferred into 2-l flask containing 1000 cm³ of the nutrient with addition of 20 mg of substrate. After six-days shaking, cells were centrifugated (2700 r.p.m.) to give 7 cm³ of the cell suspension.

2. *Immobilization by the* MOSBACH [3] *Method*

30 cm³ of phosphate buffer were added to 30 cm³ of centrifugated cells of *R. mucilaginosa* and temperature was brought to 288 K. To this suspension, 90 cm³ of phosphate buffer containing 14.2 g of acrylamide and 0.74 g of N,N-methylene-bis-acrylamide were added. After mixing, 0.16 g of ammonium persulfate dissolved in 1 cm³ of water and two drops of TEMED were added. The block obtained was rubbed through a sieve (mesh ⌀ = 0.6 mm) and decanted ten times at intervals of 5 min. using sterile phosphate buffer. So immobilized cells were principally resistant to infection in their mass but they were additionally protected by addition of an antibiotic (see points 3 and 4).

* The strain originated from the collection of Prof. A. NESPIAK, Medical Academy, Wrocław.

3. *Transformation Course*

100—150 mg of substrate dissolved in 2—5 cm^3 of acetone were introduced into 1 l of sterile phosphate buffer with addition of 200 mg of streptomycin. The solution was filtered through the layer of immobilized cells placed in the column-segment (⌀ = 3.3 cm, length 15 cm, volume ca. 130 cm^3) equipped with water jacket to maintain constant temperature (300 K). The segments were connected in series (2—4) if necessary. The rate of leakage was regulated for 0.5—1 l per 24 h. The filtrate obtained was salted out with ammonium sulfate and extracted with ethyl ether.
In the case of substrates (—)—**1**, (+)—**2**, and **4**, adsorption of substrate and product on biological material was observed. Washing the reaction column with 0.5—1 l of acetone afforded additionally ca. 40% of the reaction mixture but inactivated cells of *R. mucilaginosa*.

4. *Experimental Data and Identification of Products*

Appreciation of the yield was based on column chromatography (error up to 20%) and gas chromatography (up to 10%). Thin-layer chromatography was used for preliminary qualitative estimation of the transformation course. Gas chromatograms were performed on a Chromatograf 504 Mera Elwro apparatus with the 3% OV-17 column. IR spectra were taken on a Unicam SP-200 apparatus, UV spectra on a Specord UV VIS spectrophotometer, and MS spectra on a GC MS 2091 LKB (Sweden) spectrometer.

Transformation of (—)-menthone ((—)—1)

a. *Continuous Method*

5 l of phosphate buffer containing 100 mg of (—)-**1** and 200 mg of streptomycin in 1 l were filtered within five days through the column containing 7 cm^3 of centrifugated and then immobilized cells of *Rhodotorula mucilaginosa*. After termination of the transformation (three-fold decrease of the yield), column was washed with 1 l of acetone to afford, after salting out and extraction with ethyl ether, 350 mg (70%) of a mixture of products whose composition was determined by means of GC (Tab. 1). Column chromatography (SiO_2, hexane-ethyl ether 10:1) afforded 105 mg (30%) of (+)-neo-menthol ((+)-**1a**), $[\alpha]_{546}^{25} = +12°$ (c = 2, acetone); IR, ν_{max} (CCl_4): 3620, 3450—3560, 1450, 1380 cm^{-1}. Absolute configuration of the hydroxy group was determined by the Horeau [13] method: rotation read —0.08° (c = 1.4, benzene) proved the (S)-configuration.

b. *Submerged Method*
100 mg of (—)-menthone dissolved in 1,5 cm^3 of acetone were added to 1 l of 2% maltose nutrient containing suspension of the *R. mucilaginosa* cells and the whole was shaken for 14 days at 300 K. After salting out and extraction with ethyl ether, 80 mg of a mixture were obtained. Column chromatography (conditions as above) yielded 55 mg (ca. 70%) of (+)-neomenthol ((+)-**1a**), $[\alpha]_{546}^{26} = +14.1°$ (c = 1, acetone).

Transformation of (+)—(R)-methyl-α-campholenone ((+)—2)

1.8 l of phosphate buffer containing 100 mg of (+)-**2** and 200 mg of streptomycin in 1 l were filtered within two days through the column containing 14 cm^3 of centrifugated and immobilized cells of *R. mucilaginosa*. When the transformation was finished (three-fold decrease of the yield), column was washed with 0.5 l of acetone. After salting out and extraction, 125 mg (70%) of a product mixture were obtained, whose composition was determined by GC (Tab. 1). Column chromatography (SiO_2, pentane-ethyl ether 20:1) gave 85 mg (70%) of (+)-**2a**, $[\alpha]_{546}^{20} = +18°$ (c = 1.5, $CHCl_3$), IR, ν_{max} (CCl_4): 3633, 1470 cm^{-1}. IR spectrum and physical constants agreed with those for (+)—(R)-methyl-α-campholenol with the (S)-configuration [9].

Transformation of (+)-carvone ((+)—3)

8 l of the phosphate buffer solution containing 150 mg of (+)-**3** and 200 mg of streptomycin in 1 l were filtered through the column containing 14 cm^3 of centrifugated and immobilized cells of *R. mucilaginosa.* After 14 days of the transformation, when its yield dropped four times, filtrate was salted out and extracted with ethyl ether (substrate and product were not adsorbed on the biological material and washing of the column with acetone proved useless) to give 840 mg (70%) of a mixture of products whose composition was determined by GC (Tab. 1). The column was used once again after 10 days storing at 278 K for reduction of acetophenone (100% of reduction within further 9 days). Column chromatography (SiO_2, hexane-ethyl ether 5:1) yielded 500 mg (60%) of (—)-neoisocarveol ((—)-**3a**), $[\alpha]_{546}^{25} = -26.7°$ (c = 1.31, $CHCl_3$), IR, ν_{max} (CCl_4): 3350, 1460 cm^{-1}. The physical and spectral data agreed with those for this alcohol reported in [10].

Transformation of Eucarvone [4]

7.3 l of phosphate buffer containing 100 mg of **4** and 200 mg of streptomycin per 1 l were filtered through the column packed with 28 cm^3 of centrifugated and immobilized cells of *R. mucilaginosa.* After 8 days of the transformation, when its yield dropped four times, column was washed with 1 l of acetone. Filtrate was salted out and extracted to afford 530 mg (75%) of a product mixture, whose composition was determined by GC (Tab. 1). Column chromatography (SiO_2, pentane-ethyl ether 40:1) yielded ca. 280 mg of (—)-dihydroeucarveol ((—)-**4a**). Absolute configuration of the hydroxy group in this compound was determined by the HOREAU [13] method as (S) (the sign found —0.2°). (—)-**4a**: $[\alpha]_{546}^{21} = -1.0°$ (c = 5.67, $CHCl_3$); IR, ν_{max} (CCl_4): 3380, 1460, 1390, 1130 cm^{-1}; MS: 136 M^+ (154—18) ($C_{10}H_{18}O$), M 154/for 70 eV M for maximal signal 43 (C_2H_3O) signal M 136 was 2.87% of maximal signal for 15 eV M of maximal peak 61 (C_3H_9O) and peak with M 136 was 14.13% of maximal peak.
Ketone (—)-**5** (obtained in small amounts): $[\alpha]_{546}^{20} = -100°$ (c = 2.72, $CHCl_3$), IR, ν_{max} (film): 1705, 1470 cm^{-1}.

Acknowledgments

We thank the Polish Academy of Sciences (Grant MR.I.12) and the Polish Ministry of Science, Higher Education and Technology (Grant R.I.09) for financial support.

Received September 9, 1984

References

[1] KIESLICH, K.: Microbial Transformations of Non-Steroid Cyclic Compounds. Stuttgart: G. Thieme Publishers 1976.
[2] JIU, J.: Microbial Reactions in Prostagladin Chemistry. — In: Products from Alkanes, Cellulose and other Feedstocks. Berlin: Akademie-Verlag 1981, p. 37.

[3] MOSBACH, K., LARSON, P. O.: Biotechnol. Bioeng. **12** (1970), 19.
[4] GULAYA, V. E., ANANCHENKO, S. M., TORGOV, S. V., KOSTCHENKO, V. A., BYCHOVA, O. O.: Bioorg. Khim. **5** (1979), 768.
[5] SIEWIŃSKI, A.: Bull. Acad. Pol. Sci., Ser. Sci. Chim. **17** (1969), 475.
[6] SIEWIŃSKI, A., PECZYŃSKA-CZOCH, W., ZABŻA, A., SZEWCZUK, A.: Tetrahedron **33** (1977), 1139.
[7] NESPIAK, A., PECZYŃSKA, W., SIEWIŃSKI, A.: Arch. Immunol. Ther. Exp. **17** (1969), 433.
[8] PECZYŃSKA-CZOCH, W., SIEWIŃSKI, A., SZEWCZUK, A.: Arch. Immunol. Ther. Exp. **27** (1979), 441.
[9] SIEWIŃSKI, A., DMOCHOWSKA-GŁADYSZ, J., KOŁEK, T., ZABŻA, A., DERDZIŃSKI, K.: Tetrahedron **35** (1979), 1409.
[10] MIRONOWICZ, A., RACZKOWSKA, J., SIEWIŃSKI, A., SZYKUŁA, J., ZABŻA, A.: Pol. J. Chem. **56** (1982), 735.
[11] MIRONOWICZ, A., SIEWIŃSKI, A.: Communication on 9th Conference on Isoprenoids, Prague, September 1981.
[12] DMOCHOWSKA, J., SIEWIŃSKI, A., NESPIAK, A.: Bull. Acad. Pol. Sci., Ser. Sci. Biol. **17** (1968), 133.
[13] HOREAU, A., KAGAN, H. B.: Tetrahedron **20** (1964), 2431.

Acta Biotechnol. 6 (1986) 2, 147–152

Autolysis of Disintegrated Cells of the Yeast Saccharomyces cerevisiae

Běhalová, B., Beran, K.

Institute of Microbiology
Czechoslovak Academy of Sciences
Prague 4, Videňská 1083

Summary

Optimum conditions for autolysis of disintegrated cells of *S. cerevisiae* are at 50–53 °C and pH 5.5; the process is terminated after 6 h. In the presence of sodium chloride (3–5%) the autolysis is complete after 5 h. The yield of autolysis of disintegrated yeast cells is about 70% of autolytic product per yeast dry weight. The product obtained after centrifugation, filtration and drying has very good sensoric and physical properties.

Introduction

When they exhaust exogenous and endogenous nutrients, yeast cells suffer from autolysis during which their own lytic enzymes degrade individual macromolecular cell components [1]. The process is stimulated by elevated temperature at which the lytic enzymes do not denaturate owing to their considerable heat stability. To facilitate cell destruction, the autolysis is sometimes carried out under the addition of low concentrations of organic solvents [2, 3] which attack the lipid moiety of cell membranes, or sodium chloride [4].

An important role in autolysis of yeast cells is played by cell proteases. Yeast cells contain a complex system of intracellular proteases and peptidases and their specific protease inhibitors [5]. Proteolytic enzymes are localized in vacuoles [6], proteolytic inhibitors in the cytoplasm [7]. During disintegration the spatial separation of proteases and inhibitors is abolished and inactive complexes are formed. According to the conditions of autolysis the protease inhibitors are subsequently digested and the proteolytic enzymes attain relatively rapid their maximum activity [8, 9]. The liberation of lytic enzymes which takes several hours at an elevated temperature of autolysis in nondisintegrated cells, occurs during disintegration within several minutes and the time of autolysis is substantially shortened.

Our study of mechanism of autolysis of disintegrated cells of *S. cerevisiae* based on determination of proteolytic activity established optimum conditions of autolysis [9]. The present paper concerns the analysis of selected autolytic procedures and evaluation of quality of the resulting autolysate.

Material and Methods

The experiments were done on pressed baker's yeast. Cells were disintegrated on a mechanical disintegrated MBÚ ČSAV [10] for 20 min. The volumes of yeast suspension and lead-free glass beads 0.45—0.50 mm in diameter were in a ratio of 1:1.7, the total volume of the disintegrator was 1.2 l. Disintegration in a larger volume was done in a horizontal bead disintegrator Dyno-Mühle KD5 (Maschinenfabrik Basel, Switzerland).
Autolysis was performed with 6—8% suspensions of disintegrated yeast cells. At 1-h intervals the content of α-amino nitrogen was determined directly in suspension samples. The samples were boiled for 5 min, centrifuged, and the supernatant was used to determine dry weight and total and reducing saccharides. After complete autolysis the autolysate was freed of cell membranes and walls by 30-min centrifugation at 3500 g. The centrifuged autolysate was boiled for 10 min with charcoal and filtered through plates with kieselguhr. After an addition of NaCl it was dried in a vacuum evaporator NIRO Atomizer, type LTD (Copenhagen, Denmark) at outlet temperature of 95—97 °C.
Dry weight was determined gravimetrically on plates with sand. Protein concentration was measured according to LOWRY et al. [11]. Proteolytic activity was measured by the rate of degradation of azocasein [12, 13]. Concentration of α-amino nitrogen was determined by potentiometric titration with formaldehyde, total nitrogen was assayed in the ROY-MARGHAM apparatus after digestion [14]. Total saccharides were determined by the anthrone reaction [15], reducing saccharides according to SOMOGYI and NELSON [16], NaCl argentometrically, polymeric RNA spectrophotometrically [17], and amino acids on an amino acid analyser AAA Hd 1200 E.

Results

Time Course of Autolytic Processes

We studied in detail the proteolytic activity in autolysing suspension of disintegrated yeast cells, which is a measure of the rate of the autolytic process [9] and came to the following conclusions:

A) Optimum temperature for achieving maximum activity of proteolytic enzymes (after 7—10 h) is 40 °C.
B) Optimum temperature for rapid activation of proteolytic enzymes is 50—53 °C. Maximum activity is attained after 4—6 h but the activity of proteases does not rise to the level determined at 40 °C.
C) Sodium chloride at 40 °C inhibits activation of proteases whereas at 50 °C it accelerates the process.
D) Activation of proteases is best at pH 5—6.
Based on these results, three basic variants of autolytic conditions were chosen for further study:

1) Autolysis at 40 °C with a temperature rise to 50—53 °C after 7 h
2) Autolysis at 50 °C
3) Autolysis at 50 °C in the presence of 3—5% NaCl

The pH was 5—6 in all cases. The temperature change from 40 °C to 50 °C in variant 1, after a maximum proteolytic activity had been attained, served for acceleration of the process. The time course of autolytic processes is illustrated in Figs. 1—3.
At 40 °C the autolysis is slower than at 50 °C. The temperature change in the 7th hour causes a change in the values of protein and amino nitrogen levels, probably due to activation of other components of the proteolytic apparatus. The content of total and reducing saccharides in the autolysate at 50 °C rises up to the 6th hour when the process

is practically finished. At 40 °C the time course of increase in total sugars is similar as during autolysis at 50 °C. On the other hand, reducing sugars increase for only 2—3 h, then drop from 8% to 0.8% per autolysate dry weight. Hence at 40 °C the final autolysate contains practically no reducing sugar whereas at 50 °C the sum of reducing and polysaccharide compounds accounts for a considerable portion of the dry weight.

Autolysis Yields

Individual variants differ in the time necessary for completion of autolysis. Variant 1 (40 °C with change to 50 °C) was terminated after 10 h, variant 2 (50 °C) after 6 h and variant 3 (50 °C, addition of NaCl) after 5 h. Mass and nitrogen balance was determined

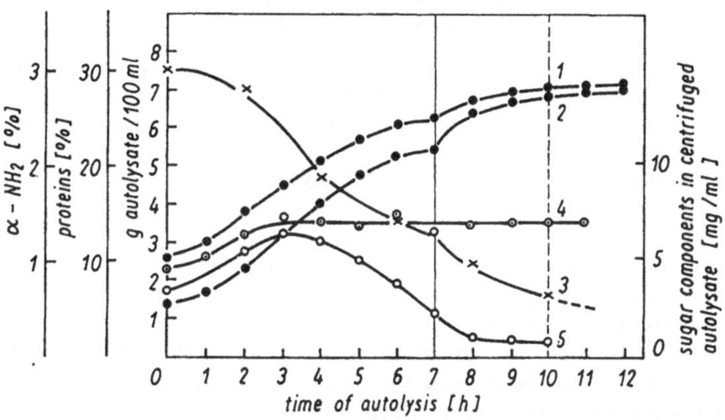

Fig. 1. Time curse of autolysis of disintegrated yeast cells (*Saccharomyces cerevisiae*) at 40 °C, with a temperature shift to 50 °C in the 7th hour.
1 — yield (g autolysate dry weight/100 ml), 2-α-aminonitrogen, 3 — proteins, 4 — total sugars, 5 — reducing sugars
Dashed line — termination of autolysis. Full vertical line — change of temperature.

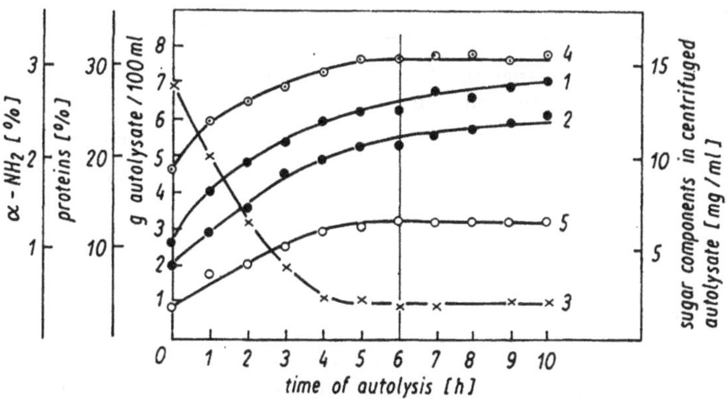

Fig. 2. Time course of autolysis of disintegrated cells of *Saccharomyces cerevisiae* at 50 °C.
1 — yield (g autolysate dry weight/100 ml), 2 — α-aminonitrogen, 3 — proteins, 4 — total sugars, 5 — reducing sugars
Full vertical line — termination of autolysis.

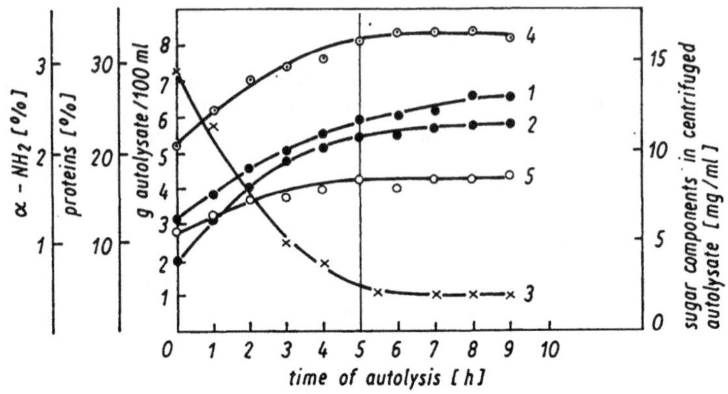

Fig. 3. Time course of autolysis of disintegrated cells of *Saccharomyces cerevisiae* at 50 °C in the presence of 4.5% NaCl.
1 — yield (g autolysate dry weight/100 ml), 2 — α-amino nitrogen, 3 — proteins, 4 — total sugars, 5 — reducing sugars.
Full vertical line — termination of autolysis.

at three selected standard conditions. Table 1 compared the above variants with the autolysis of nondisintegrated yeast cells with the addition of 2% chloroform; the regime of autolysis of disintegrated cells is seen to have no effect on the product (about 70% of cell dry weight) or on the yield of the nitrogen fraction (91—93%). With nondisintegrated cells a more than double time of autolysis results in product yield of a more 50% and nitrogen substances yield of some 70%.

Table 1. Comparison of autolytic processes used with disintegrated cells of *Saccharomyces cerevisiae*

Conditions				Yield [%]	
Temperature [°C]	Note	pH	Time [h]	Dry weight	Nitrogen substances
40°	Change to 52 °C in the 7th hour	5.5	10	69	93
52	—	5.5	7	68	91
52	Activation with 4% NaCl	5.5	6	71	91
50	Without disintegration	5.5	24	50	70

Analytical Evaluation of Product

Autolytic processes at 40 °C and 50 °C should be taken as two different processes which differ in the set of enzymes participating in the autolysis; this is reflected in the different time of autolysis, chemical composition of the autolysates and their organoleptic properties. Products produced at 50 °C are fully acceptable as to their flavour and aroma whereas products obtained at 40 °C have an unpleasant sour flavour.
The analysis was performed on five batches of autolysate produced at 50 °C. The preparation, characteristics and analytical data for these batches are given in Table 2. The

Table 2. Analysis of autolytic products of disintegrated cells of *Saccharomyces cerevisiae*

Sample no.		1	2	3	4	5	ALBRECHT DEIN-DOERFER (1966)	BALDES (1971)
Water content	[%]	8.3	6.9	8.3	7.3	6.9	3.5	18.6
NaCl	[%]	41.8	49.5	44.0	43.5	43.0	46.5	16.0
Nitrogen	[%]	7.95	6.7	7.0	7.6	7.1	7.3	7.0
Total sugars	[%]	13.4	23.8	21.8	20.8	10.5	—	—
Reducing sugars	[%]	8.5	14.9	15.0	9.1	3.0	—	—
Polymeric RNA	[%]	0.11	0.14	0.16	0.09	0.16	—	—
Crude protein	[%]	49.3	42.8	43.7	47.8	44.7	45.6	43.8
α-Amino nitrogen	[%]	3.7	3.3	3.0	3.9	4.0	—	—
Ratio α-NH_2/N	[%]	46.4	49	43	51	56	—	—

Sample 1—51 °C, 6.5 h, +3% NaCl; sample 2—53 °C, 6.0 h, +3% NaCl; sample 3—50—52 °C, 5.5 h; sample 4—50—52 °C, 6.5 h; sample 5—47—49 °C, 7.0 h, semi-pilot plant.

Table 3. Amino acids in different types of autolysates from disintegrated cells of *Saccharomyces cerevisiae*

Amino acid		Type of autolysis			
		50 °C (+ NaCl)	50 °C	48 °C	50 °C + chloroform without disintegration
Phenylalanine	[%]	1.80	0.37	3.40	1.58
Lysine	[%]	4.20	3.60	2.30	5.26
Arginine	[%]	2.66	2.55	stopy	1.41
Histidine	[%]	0.43	1.50	stopy	1.03
Asparagine	[%]	4.88	5.10	4.55	4.53
Threonine	[%]	1.98	2.40	2.55	2.03
Serine	[%]	2.09	2.37	2.55	2.15
Glutamic acid	[%]	4.25	8.23	8.91	9.72
Proline	[%]	0.90	1.84	2.37	1.57
Glycine	[%]	2.02	2.40	2.50	2.46
Alanine	[%]	3.97	3.70	4.48	4.59
Cysteine	[%]	—	—	—	—
Valine	[%]	2.80	3.12	3.87	2.54
Methionine	[%]	0.03	0.55	0.87	0.65
Isoleucine	[%]	2.46	2.83	3.19	1.83
Leucine	[%]	3.66	3.28	4.65	2.57
Tyrosine	[%]	2.02	0.67	1.59	1.19
Total amino acids per dry weight of yeast fraction	[%]	41.1	44.4	44.8	45.1

samples complied generally with required quality standards. Out of total nitrogen a high proportion is due to α-amino nitrogen. Undesirable ribonucleic acids are mostly degraded by the autolysis. Autolysate from a semipilot plant batch has the lowest content of both total and reducing sugars since the autolysis was performed at a slightly lower temperature (47—49 °C), and entailed very probably some degradation of reducing sugars by the enzyme apparatus which is otherwise inhibited at higher temperatures.
The content of amino acids in the samples and their comparison with autolysate produced in the conventional way is given in Table 3. No substantial differences were observed in the total content of amino acids per dry weight of the yeast fraction.

Conclusion

Autolysis of disintegrated cells of Saccharomyces cerevisiae proceeds at a much higher rate and with higher yields as compared with the autolysis of intact cells.
Under optimum conditions, i.e. at 50—53 °C, pH 5.5 and the presence of NaCl the degradation of proteins is complete within 5 h. The original content of proteins, which form 25—30% of the yeast dry weight, decreases during autolysis to 4—6%, corresponding to a 80—85% degradation of the proteins. The yield of autolysis of disintegrated yeast cells is about 70% of the product. The autolysate obtained after centrifugation, filtration and drying has very good sensoric and physical properties. Without the addition of NaCl the process is somewhat slower but the dry weight yield, nitrogen yield and sensoric properties are the same.

Received December 11, 1984

References

[1] Suomalainen, H.: 11. ICC-Meeting. Vienna 1974. P. 67.
[2] Peppler, H. J.: The Yeast. Eds.: A. H. Rose, J. S. Harrison. Vol. 3. New York, London: Academic Press, 1970, p. 442.
[3] Reed, G., Peppler, H. J.: Yeast technology. Westport, CT: AVI Publishing Company, 1973.
[4] Sugimoto, H.: J. Food Sci. **39** (1974), 939.
[5] Holzer, H.: J. Biochem. (Tokyo) **79** (1976), 39P.
[6] Matile, P., Wiemken, A.: Arch. Mikrobiol. **56** (1967), 148.
[7] Holzer, H., Saheki, T.: Tokyo J. Exp. Clin. Med. **1** (1976), 115.
[8] Saheki, T., Holzer, H.: Biochem. Biophys. Acta **384** (1975), 203.
[9] Běhalová, B., Beran, K.: Folia Microbiol. **24** (1979), 455.
[10] Řeháček, J., Beran, K., Bičik, V.: Appl. Microbiol. **17** (1969), 462.
[11] Lowry, O. H., Rosebrough, N. J., Farr, A. L., Randall, R. J.: J. Biol. Chem. **193** (1951), 265.
[12] Fabián, J.: Folia Microbiol. **15** (1970), 160.
[13] Hasen, G. G., Hanse, J. A., Hubicky, J. A.: Ann. N.Y. Acad. Sci. **130** (1965), 761.
[14] Margham, R.: Biochem. J. **36** (1942), 790.
[15] Morris, D. L.: Science **107** (1948), 254.
[16] Somogyi, M.: J. Biol. Chem. **195** (1952), 19.
[17] Schmidt, G., Thannhauser, S. J.: J. Biol. Chem. **161** (1945), 83.
[18] Albrecht, J., Deindoerfer, F.: Food Eng. **28** (1966), 92.
[19] Baldes, W.: Z. Lebensm.-Unters.-Forsch. **145** (1971), 149.

Acta Biotechnol. 6 (1986) 2, 153—160

Xylan-Containing Supports Used to Separate Xylanase from the Cellulases of Aspergillus terreus F-413

PASZCZYŃSKI, A.[1], ROGALSKI, J.[1], SZCZODRAK, J.[2], DAWIDOWICZ, A.[3]

[1] M. Curie-Skłodowska University
Department of Biochemistry
M. Curie-Skłodowska Sq. 3, 20-031 Lublin, Poland
[2] M. Curie-Skłodowska University
Department of Applied Microbiology
M. Curie-Skłodowska Sq. 3, 20-031 Lublin, Poland
[3] M. Curie-Skłodowska University
Department of Chemical Physics
M. Curie-Skłodowska Sq. 3, 20-031 Lublin, Poland

Summary

The separation of xylanase from cellulolytic enzymes of *A. terreus* F-413 by affinity chromatography on xylan-containing supports was investigated. Xylanase purified over tenfold was obtained after column chromatography on xylan bound to controlled porous glass. The molecular weight of the purified enzyme has been found to be 140000 daltons, and it is homogeneous in polyacrylamide gel electrophoresis. Purified xylanase also showed residual cellulolytic activity (perhaps cross-specificity) with cellulosic substrates.

Introduction

Hemicellulolytic enzymes have been examined from a wide variety of microorganisms, predominantly fungi, and occasionally bacteria and protozoa [1—3]. The enzymes of fungal origin are of specific interest, since so many of these organisms depend on efficient extracellular cellulolytic and hemicellulolytic systems for their existence.

Xylan hemicelluloses are hydrolyzed by two enzyme systems. Endo-xylanase (EC. 3.2.1.8), (1,4-β-D-xylanohydrolase), an endo-acting enzyme which produces initial depolimerization of the substrate, and saccharifying β-xylosidase (EC. 3.2.1.37), (β-D-xyloside xylohydrolase), an exo-splitting enzyme which cleaves the same substrate or xylooligosaccharides and xylobiose to xylose [4, 5].

The xylanase activity in genus *Aspergillus* was investigated from 1955 [6], IWAMOTO et al. [7] reported two forms of xylanases produced by *A. niger*. FUKUMOTO et al. [8] found an isolate of *A. niger* possessing three forms of this enzyme. JOHN and coworkers [9] described five different xylanases, whereas FREDERICK et al. [10] purified four fractions of this enzyme from this organism. The number of xylanase forms probably depends on the kind of the strain within the same species, because TAVOBILOV et al. [11] purified only one xylanase from *A. niger* 15.

The results of our earlier experiments on the composition of the cellulolytic complex of the strain *A. terreus* F-413 have shown that, besides cellulases, it demonstrated the ability to decompose xylan as well as a small lignolytic activity [12].

Xylanolytic activity of the mould *A. terreus* is often very closely associated with its cellulolytic complex [13]. Therefore, it is very difficult to separate several enzymes of the xylanase system from their cellulolytic counterparts.

In the present report we have described a simple affinity chromatography technique of xylanase separation from the cellulases complex isolated from the mould *A. terreus* F-413.
New supports: xylan-AH-Sepharose and xylan-controlled porous glass were synthetized for this purpose. Before xylan bonding the glass support was in addition thermally treated and hydroxylated to form a greater amount of B—OH groups on CPG surface [14—18]. As it was shown, such a procedure allows to obtain packings with better chromatographic properties [17, 18].

Materials and Methods

Enzymes Source and Activity

The strain *Aspergillus terreus* F-413 from the Czechoslovak Collection of Microorganisms in Brno was maintained at 4 °C on agar slants of the medium of SAUNDERS et al. [19] with the addition of 1% swollen phosphoric cellulose and 0.05% neopeptone. Liquid SAUNDERS medium with the addition of 0.05% neopeptone and 1% powdered cellucotton (170—200 mesh) was used for the submerged culture. The 10 l fermenter filled with 6 l of the medium was sterilized (0.1 MPa for 1 h) and inoculated with 500 ml of 96-hour old shake culture. Growth conditions: aeration — 1 l $\times$ min^{-1} $\times$ 1 litre medium^{-1}, temperature 30 °C, stirring speed 350—400 rpm., volume 6 l.
Crude cellulase preparation containing xylanase activity were precipitated from the postculture liquid of *A. terreus* F-413 with ethanol containing 3% acetone [20]. This preparation contained several specific activities of enzymes (units/mg protein): xylanase (5.1 U) — 1 μmol of reducing equivalent calculated as xylose liberated from xylan in 1 min at 55 °C was assumed as activity unit; saccharifying cellulase activity (0.19 U) — one unit is expressed as 1 μmol of reducing equivalent, calculated as glucose liberated from microgranular cellulose CC-41 per 1 min at 50 °C; aryl-β-glucosidase (0.28 U) — 1 μmol of reducing equivalent calculated as glucose produced from salicin as substrate in 1 min at 50 °C was assumed as activity unit; β-glucosidase (cellobiase) (0.17 U) — one unit of enzymatic activity is defined as 1 μmol of glucose formed from cellobiose per 1 min at 50 °C. The preparation was desalted on a Sephadex G-25 column (26 $\times$ 9 cm), lyophilized and used as crude material in affinity chromatography experiments.

Preparation of Glass Beads

The porous glass support (CPG) was prepared from Vycor glass [21, 22] consisting of 7 mol % Na_2O, 23 mol % B_2O_3 and 70 mol % SiO_2.
A 102—120 μm fraction of this glass was heated at 600 °C for 25 h. During the heating period the liquation process took place [21, 22] in the glass and as a result, a new phase very rich in alkali borate was formed in the continuous silica network. After the liquation process the heated glass was leached, first by hot 5N H_2SO_4 and then by 0.5 N NaOH at room temperature. After each leaching procedure the prepared material was carefully washed to neutrality and dried at 120 °C. Details of the applied procedure were previously described in our earlier papers [16, 17, 23, 24].
Porous glass resulting from extracted sodium-boro-silicate glass still contained residual boron oxide (3%) and sodium oxide (~ 0.5%) in the silica network of the glass. In order to enrich the glass surface with boron atoms, the obtained controlled porous glass (CPG) was heated at 700 °C for 5 h [15, 16, 18]. The thermal treatment of CPG causes not only the diffusion of boron atoms from the bulk to the surface [14—17] but also leads to a surface dehydroxylation [15—17]. The glass surface enriched with boron was rehydroxylated by the procedure described earlier [17].

Activation of the Supports

Glass supports were activated by γ-aminopropyltriethoxysilane (γ-APTES) according to ROBINSON et al. [25]. AH-Sepharose was obtained from Pharmacia, Uppsala, Sweden. Xylan was activated by the cyanogen bromide method [26]. The supports were coupled to the xylan by the method of LAPPI et al. [27].

Conditions of Xylanase Purification

The crude enzymes dissolved in 0.0025 M Na-citrate buffer (pH 5.0, 40 mg/3 ml) were poured on two kinds of columns (1 cm × 6 cm) xylan-AH-Sepharose or xylan-controlled pore glass which were first equilibrated and eluted with the same buffer. Fractions of 2.5 ml were collected at the elution rate of 17 ml/h. The concentration of the Na-citrate buffer was changed from 0.0025 M to 0.1 M and pH from 5.0 to 3.0. Fractionation of proteins was obtained by a stepwise increase of the ionic strength and hydrogen ions concentration. Protein was determined by LOWRY et al. [28] method. Both columns were regenerated with 6 M urea.

Enzyme Assays

The activities of xylanase and cellulases (aryl-β-glucosidase, β-glucosidase) were assayed according to the methods described earlier [12, 20]. Saccharifying cellulase activity was determined using microgranular cellulose CC-41 (WHATMAN) as a substrate. The reaction mixture contained 0.5 ml of 1% cellulose suspension and 0.5 ml of appropriately diluted enzyme in 0.05 M Na-citrate buffer, pH 4.8. After 30 min of incubation at 50 °C the reaction mixture was centrifuged, and in the supernatant reducing sugars were determined by the method of SOMOGYI-NELSON [29, 30].

Disc Gel Electrophoresis

Protein electrophoresis was performed in a discontinuous electrolyte system after LAEMMLI [31]. For this purpose 6% polyacrylamide gels were used. According to electrophoresis the proteins bands were stained for 0.5 h with 0.15% Coomassie brillant blue R-250 dissolved in methanol: acetic acid : glycerol : water (16:2:1:23, by vol.) and destained in methanol : acetic acid : glycerol : water (8:2:1:29, by vol.). For determination of the xylanase molecular weight HMW Calibration Kit Proteins-Lot 2055, Pharmacia (Uppsala) was used.

Results and Discussion

In the studies presented an attempt was made to separate the activity of xylanase from the complex of cellulases of the fungus *A. terreus* F-413 by means of affinity chromatography. For this purpose two new supports, xylan-AH-Sepharose and xylan-controlled porous glass, were synthetized.

Fig. 1 presents the fractionation of crude enzymatic preparations of *A. terreus* F-413 on xylan-AH-Sepharose. Three distinct protein fractions (absorbance at 280 nm) were obtained as a result of the separation of the material studied. The first fraction, I, (not adsorbed on the column) possessed only a low xylanase activity. The protein fraction II (eluted with 0.01 M citrate buffer, pH 3.0) showed the highest xylanase activity (the specific activity of the enzyme reached a value of about 9 U). No activity with regard to xylan decomposition was found in the third fraction, III, (eluted with 0.1 M buffer).

During chromatography of the crude cellulase preparations on xylan-controlled pore glass three protein fractions were also obtained, and in two of them (fractions I and II) xylanase was found (Fig. 2). As in the case of fractionating the material studied on xylan-AH-Sepharose, the second peak possessed the highest specific activity of xylanase (55.5 U). However, attention is draw by the fact that the activity was over sixfold higher than that obtained by using xylan-AH-Sepharose. Thus the use of xylan immobilized on glass beads permitted to obtain a considerably higher purification degree of xylanase than in the case of chromatography on xylan-AH-Sepharose.

In order to find the so-called "cross-specificity" of the purified xylanase preparations, the second fraction obtained on both supports and containing the highest specific activities of xylanase was examined for cellulolytic activity. The results are presented in Table 1, in which the data concerning the activity of the enzymes contained in the crude

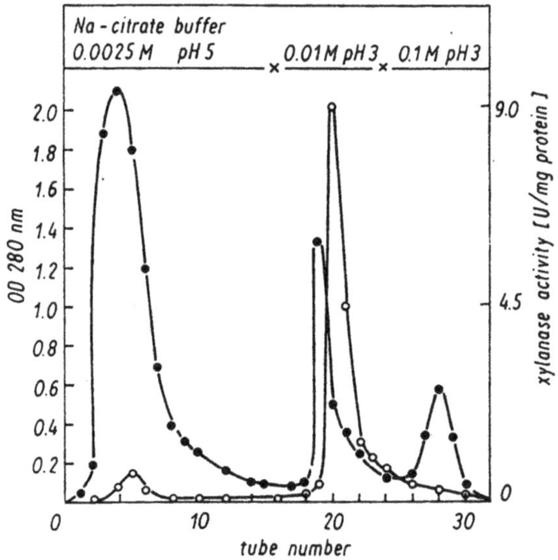

Fig. 1. Affinity chromatography of xylanase on xylan-AH-Sepharose 4B column. Conditions as described in Materials and Methods. The column was equilibrated with 0.0025 M sodium citrate buffer pH 5.0. ○— xylanase activity, ●— absorbance at 280 nm, temperature 5—10 °C

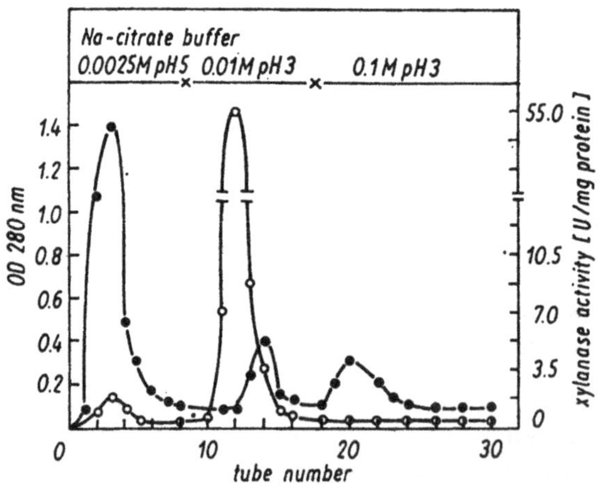

Fig. 2. Affinity chromatography of xylanase on xylan-CPG column. The experimental conditions were the same as described in the text.
○— xylanase activity, ●— absorbance at 280 nm, temperature 5—10 °C

preparation have additionally been summarized. From the results it appears that the xylanase preparations studied, despite their purification, still possessed small activities of cellulases (saccharifying activity, β-glucosidase and aryl-β-glucosidase). An analysis of the data concerning the specific activities of xylanase contained in the crude and purified preparations on the supports containing xylan has accounted for over a tenfold purification of this enzyme on a column with xylan-controlled pore glass and for only

Table 1. Cross-specificity of xylanase preparations

Enzymes	Specific activity [%]		
	Crude	After chromatography on xylan-controlled pore glass	After chromatography on xylan-AH-Sepharose
Xylanase	100.00	1088.00	176.47
Saccharifying cellulase	4.57	0.57	0.92
β-glucosidase	3.91	0.44	2.00
aryl-β-glucosidase	6.52	0.80	3.75

The values in the table represent the amount of sugar liberated from the substrates used for studies of enzymatic activities relative to that from xylan.

its twofold purification on a column with xylan-AH-Sepharose (Table 1). A similar purification degree of the enzyme (tenfold) was obtained in their studies on *Gliocladium virens* by TAKAHASHI and KUTSUMI [32]. However, in contrast to our results, the conventional chromatography techniques used by these authors gave this effect only after 8 stages of purification.

The second fraction obtained on both supports and containing the highest activities of xylanase was subjected to electrophoresis on polyacrylamide gel (Fig. 3). The electrophoretic studies have permitted to determine the molecular weight of the enzyme for about 140000 D. Pure xylanase preparation was homogeneous in electrophoresis only after purification on xylan-controlled pore glass, but the enzyme preparation purified on xylan-AH-Sepharose contained an additional protein band besides the main xylanase fraction.

Both columns used six times did not change their capacity. After regeneration with 6 M urea trace amounts of reducing sugars were found in the effluent. As in the studies

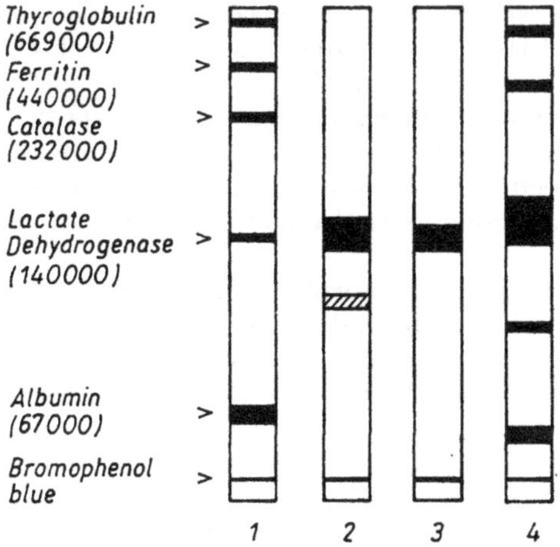

Fig. 3. Polyacrylamide gel electrophoresis of the purified and crude xylanase preparations. Gel electrophoresis was performed by the procedure as described in Materials and Methods.
1 markers, 2 after xylan-AH-Sepharose chromatography, 3 after xylan-controlled pore glass chromatography, 4 crude. Electrophoresis was conducted at 3 mA per tubes (0.6 ⌀ × 9 cm)

on glucoamylase the support with AH-Sepharose showed a smaller selectivity than the glass support [33]. As it was mentioned before, glasses of controlled porosity contain $\sim 3\%$ B_2O_3 in their silica network. A characteristic feature of CPG is the possibility of diffusion of boron atoms remaining in the silica network from the bulk to the surface [14, 15]. In the previous papers [17, 18] it was shown that the structure of films bounded to CPG surface enriched with boron is denser and more rigid than analogous one bonded to normal CPG or silica gel. Moreover, such materials, where radical or affinant is bonded, with the surface via B—OH group, interact more specifically with chromatographed molecules [17, 18, 34, 35].

Although affinity chromatography was used, a complete separation of xylanase activity from the complex of cellulases of *A. terreus* F-413 was not attained. This may account for a very close relationship between the two groups of enzymatic activities, or for the occurrence of the so called "cross-activities". A considerable doubt existed for some time whether cellulases and xylanases were separate enzymes. The likelihood of significant cross-specificity and the inability to demonstrate enzyme heterogeneity in purified cellulase preparations which also possessed xylanase activity argued for a single multifunctional enzyme. TODA et al. [36], from studies with highly purified cellulase components from *Trichoderma viride*, concluded that at least one of the two cellulase components examined also hydrolyzed xylan as an alternative substrate. In 1971, ERIKSSON and PETTERSSON [37] obtained and characterized a highly purified endo-xylanase from the rot fungus *Stereum sanguinolentum*, the same organism from which BJÖRNDAL and ERIKSSON [38] had previously obtained a highly purified cellulase. Earlier, HRAZDINA and NEUKOM [39] had succeeded to separate a "practically pure xylanase" from a commercial crude cellulase preparation which still retained traces of cellulolytic activity. JOHN et al. [9] purified to homogeneity five different xylanases from the culture medium of *Aspergillus niger*, among which two showed a relatively broad substrate specificity and were able to degrade crystalline cellulose. Also TAKAHASHI and KUTSUMI [32] purified endo-1,4-β-xylanase from the culture broth of *Gliocladium virens* to a crystalline form, which also showed a weak activity toward CM-cellulose, cellulose powder or starch. Latter, SREENATH and JOSEPH [40] fractionating extracellular xylan hydrolases of a strain of *Streptomyces exofoliatus* obtained three major fractions of xylanases. One of the endo-enzymes was specific for xylan, whereas the other could also attact cellulose, inulin and pectin.

These observations are in agreement with our experiments with respect to the mould *A. terreus* F-413, in which a close relationship between the two groups of enzymes as well as the occurrence of "cross-specificity" between them have been found, although a highly specific method of affinity chromatography has been used for purification of these enzymes.

Xylanase has been often recognized as a polymorphic enzyme. For example, one to five different fractions of this enzyme have been described within the species *A. niger* [8, 9, 11]. REILLY [41] supposes that the occurrence of so numerous and different forms of xylanase in this fungus may result from a modified action of proteases contained in the preparations studied on this enzyme. As it appears from our earlier studies in a 6-day culture of *A. terreus* F-413 on cellucotton, no measurable activities of proteases have been found in the postculture liquid [12]. Therefore, it is likely that the purified xylanase preparation was as mentioned above, homogeneous.

The affinity chromatography on xylan supports (particularly xylan-controlled pore glass) synthetized by us permits to separate selectively xylanase from crude enzymatic preparations of the fungus *A. terreus* F-413. This method, in contrast to other conventional chromatographic techniques, constitutes a single-stage process which is very efficient and permits to obtain purified xylanase preparations in a short time.

Acknowledgments

The authors wish to thank Doc. Dr. J. ŁOBARZEWSKI from the Department of Biochemistry, Maria Curie-Skłodowska University, Lublin for the discussion of results presented in this paper. This work was supported by the Research Programme MR-I-14 and R.1.9/05.05.

Received October 31, 1984

References

[1] KISIELEWSKA, E.: Ann. Univ. M. Curie-Skłodowska, Sec. C. **32** (1977), 193.
[2] KUNERT, J., ZEMEK, J., AUGUSTIN, J., KUNIAK, L., LÝSEK, H.: Biológia **37** (1982), 291.
[3] ESTEBAN, R., VILLANUEVA, J. R., VILLA, T. G.: Can. J. Microbiol. **28** (1982), 733.
[4] DEKKER, R. F. H., RICHARDS, G. M.: Adv. Carbohydr. Chem. Biochem. **32** (1976), 277.
[5] DEKKER, R. F. H.: Polysaccharides in Food. Eds.: J. M. V. BLANSHARD and J. R. MITCHELL. London—Boston: Butterworths, 1979, p. 93.
[6] WHISTLER, R. L., MASAK, E. Jr.: J. Am. Chem. Soc. **77** (1955), 1241.
[7] IWAMOTO, T., SASAKI, T., INAOKA, M.: Mem. Ehime Univ. **17** (1972), 185.
[8] FUKUMOTO, J., TSUJISAKA, Y., TAKENISHI, S.: Nippon Nogei Kagaku Kaishi **44** (1970), 447.
[9] JOHN, M., SCHMIDT, B., SCHMIDT, J.: Can. J. Biochem. **57** (1979), 125.
[10] FREDERICK, J. R., FRATZKE, A. R., OGUNTIMEIN, G. B., FREDERICK, M. M., REILLY, P. J.: Ruth Chem. Eng. Res. Symp. (Pap.), Paper No. 7, Iowa (1977), 27 pp.
[11] TAVOBILOV, J. M., GORBACHYOVA, I. V., RODIONOVA, N. A., BEZBORODOV, A. M.: Prikl. Biokhim. Mikrobiol. **17** (1981), 436.
[12] SZCZODRAK, J., TROJANOWSKI, J., ILCZUK, Z, GINALSKA, G.: Acta Microbiol. Pol. **31** (1982), 257.
[13] LOGINOVA, L. G., GUZHOVA, E. P., ISMAILOVA, D. YU., BURDENKO, L. G.: Prikl. Biokhim. Mikrobiol. **14** (1978), 485.
[14] LOW, M. J. D., RAMASUBRANIAN, N.: J. Phys. Chem. **71** (1967), 3077.
[15] KIRUTENKO, V. M., KISELOV, A. V., LIGIN, V. I., SHTCHEPALIN, K. L.: Kinetika i Kataliz **15** (1974), 15.
[16] DAWIDOWICZ, A. L., CHOMA, I.: Materials Chem. and Phys. **8** (1983), 323.
[17] DAWIDOWICZ, A. L., RAYSS, J., SUPRYNOWICZ, Z.: Chromatographia **17** (1983), 157.
[18] RAYSS, J., DAWIDOWICZ, A. L., SUPRYNOWICZ, Z., BUSZEWSKI, B.: Chromatographia **17** (1983), 437.
[19] SAUNDERS, P. R., SIU, R. G. H., GENEST, N.: J. Biol. Chem. **174** (1948), 697.
[20] ROGALSKI, J., SZCZODRAK, J., ILCZUK, Z.: Acta Microbiol. Pol. **32** (1983), 363.
[21] HALLER, W. J.: J. Chem. Phys. **42** (1965), 686.
[22] VOLF, M. B.: Technical Glasses. London: Sir I. PITMAN and sons Ltd., 1961, p. 176.
[23] DAWIDOWICZ, A. L., WAKSMUNDZKI, A., DERYŁO, A.: Chem. Anal. **24** (1979), 811.
[24] ŁOBARZEWSKI, J., DAWIDOWICZ, A. L.: PL-P 241—749 (1983).
[25] ROBINSON, P. J., DUNNILL, P., LILLY, M. D.: Biochim. Biophys. Acta **242** (1971), 659.
[26] LOWE, C. R., DEAN, P. D. G.: Affinity Chromatography. London, New York, Sydney, Toronto: J. WILLEY and Sons, 1974, p. 250.
[27] LAPPI, D. A., STOLZENBACH, F. E., KAPLAN, N. O., KAMEN, M. D.: Biochem. Biophys. Res. Commun. **69** (1976), 878.
[28] LOWRY, J. O. H., ROSENBROUGH, N. J., FARR, A. L., RANDALL, R. J.: J. Biol. Chem. **193** (1951), 265.
[29] SOMOGYI, M.: J. Biol. Chem. **160** (1945), 61.
[30] NELSON, N.: J. Biol. Chem. **153** (1944), 375.
[31] LAEMMLI, U. K.: Nature **227** (1970), 680.
[32] TAKAHASHI, M., KUTSUMI, S.: J. Ferment. Technol. **57** (1979), 434.

[33] Paszczyński, A., Miedziak, I., Łobarzewski, J., Kochmańska, J., Trojanowski, J.: FEBS Letters **149** (1982), 63.
[34] Dawidowicz, A. L., Łobarzewski, J.: Chromatographia 18 (1984), 41.
[35] Dawidowicz, A. L., Łobarzewski, J.: Chromatographia 18 (1984), 389.
[36] Toda, S., Suzuki, H., Nishizawa, K.: J. Ferment. Technol. 48 (1970), 580.
[37] Eriksson, K.-E., Pettersson, B.: Int. Biodeterior. Bull. 7 (1971), 115.
[38] Björndal, H., Eriksson, K.-E.: Arch. Biochem. Biophys. **124** (1968), 149.
[39] Hrazdina, G., Neukom, H.: Biochim. Biophys. Acta **128** (1966), 402.
[40] Sreenath, H. K., Joseph, R.: Folia Microbiol. **27** (1982), 107.
[41] Reilly, P. J.: Enzyme technology in the utilization of agricultural wastes. Final report. ISU-ER1-Ames-77312. ERI Project 11348 Iowa State Univ. Ames (1977), 1—83.

Acta Biotechnol. 6 (1986) 2, 161–166

Optimal Production of Cellulolytic Enzymes and their Location in Trichoderma pseudokoningii

KALRA, M. K., SANDHU, D. K.

Department of Biology
Guru Nanak Dev University
Amritsar-1430005, India

Summary

The maximum production of cellulose enzymes (FPactivity, CMcellulase and β-glucosidase) in the culture filtrate was observed at 27 °C and pH 5.0, except for β-glucosidase, being at 6.0. The levels of extracellular enzymes in shake and stationary culture conditions was almost equal although they reached their maxima earlier in the former. The distribution of the three enzymes in extracellular, cytosol and cell debris fractions revealed most of the FPactivity and CMcellulase occurring extracellulary. More than 50% of the β-glucosidase was present in the cell debris fraction.

Introduction

Cellulolytic enzymes play an important role in the complete saccharification of cellulose, which has a considerable potential as a source of chemicals, energy and microbial protein. These enzymes have been reported from various organisms of which *Trichoderma reesei* is generally regarded as the best producer [1–3]. Many reports published earlier give in detail the nature and mode of action of these enzymes [4]. To obtain more active strains showing higher rates of hydrolysis a mutation programme combined with optimization of culture conditions has been suggested [5–7]. In a screening programme *Trichoderma pseudokoningii* was isolated from degrading *Dalbergia* bark. This strain growing on native straws could produce cellulases comparable to those of *T. reesei* QM 9414 in the culture filtrate [8]. This paper deals further with the optimal production and location of the three cellulolytic enzymes.

Materials and Methods

Organism

Trichoderma pseudokoningii (ATCC 52326) was maintained in soil at 4 °C. Fresh culturng was done on VOGEL's medium glucose agar whenever reguired.

Inoculum

A uniform suspension of spores was prepared from a six day old culture and filtered through gauze to remove mycelial debris. All operations were done under sterile conditions. A standard inoculum giving final concentration of 5×10^6 spores/ml of the medium was used throughout the study.

Enzyme Production

Twenty five ml of VOGEL's medium [9] supplemented with 1% carboxymethyl cellulose (CMC) (Sigma) was dispensed in 100 ml Erlenmeyer flasks. The flasks were autoclaved at 15 lbs pressure for 20 min. After inoculation, incubation was done at 27 °C as stationary or shake cultures. Six flasks per treatment were analysed daily for 10 days.

Initial pH of the Medium and Temperature of Incubation

To find out the optimal pH for cellulase production the growth medium was adjusted to different pH values ranging from 2—8. The inoculated cultures were incubated at 27 °C as stationary cultures. For temperature studies the medium inoculated with the test organism was incubated at 15, 22, 27, 32, 37 and 42 °C as stationary cultures. Flasks were analysed after 5 days of incubation.

Fractionation of the Samples

Each sample was filtered through a preweighed sintered glass funnel at 4 °C. The resulting culture filtrate was centrifuged at 8. 500 × g for 15 min at 4 °C and stored for further enzyme analysis and unused substrate. Mycelium was washed with 0.1 M acetate buffer at pH 5.0 and dried between folds of filter paper. The above mycelium was frozen and crushed with chilled, acid washed sand to prepare cytosol extracts and for cell debris fractions, it was macerated without sand. Both the fractions were suspended in buffer and centrifuged at 22.000 × g for 30 min at 4 °C. The supernatant was taken as the cytosol fraction while the pellet of cell debris without sand was resuspended in 5 ml of buffer and stored at —15 °C if not analysed immediately. For dry weight determination the mycelial samples were washed thoroughly with deionized water and dried at 75 °C to a constant weight.

Enzyme Assays

Filter paper activity (FPactivity) and carboxymethyl cellulase (CMcellulase) were estimated as described earlier [9], using filter paper strips and CMC as the substrates. For measuring β-glucosidase, to 0.5 ml of diluted enzyme sample was added 0.5 ml of 5 mM p-nitrophenyl-β-D-glucopyranoside (PNPG) (Sigma). The mixture was incubated at 37 °C for 30 min. The reaction was terminated by adding 4 ml of 0.2 M NaOH/glycine buffer at pH 10.6 and the absorbance read at 420 m.
Carboxymethyl cellulose in the culture filtrate was determined with anthrone reagent [10].

Enzyme Unit

One unit of enzyme activity is equivalent to 1 μmole of product released/min. Cytosol and cell debris-associated enzymes have been expressed in terms of specific activity as units/mg protein for cytosol and units/mg dry weight for cell debris. The soluble protein content was estimated by the method of LOWRY et al. [11].

Results and Discussion

Extracellular production of all the three components of cellulase as studied in other fungi [12, 13] was found to be linear with growth. Their levels in the two culture conditions were almost equal although they reached their maxima on 3rd day in shake and between 5—7 days in stationary culture followed by a steady decline (Fig. 1). The maximum enzyme yields in the stationary culture were 0.11 units for FPactivity, 0.55 units for CMcellulase and 0.097 units for β-glucosidase.

The profiles of the enzymes in cytosol and cell-debris fractions were almost similar to those obtained for extracellular enzymes (Fig. 1) but FPactivity and CMcellulase were either absent or present in residual amounts in shake cultures, thus showing better suitability of stationary conditions. On the other hand, β-glucosidase in the above fractions of shake culture was higher or equal to that present in stationary culture (Fig. 1h, i). The distribution of the three enzymes in the different fractions as observed on the day of maximum enzyme production revealed most of the FPactivity and CMcellulase occurring extracellularly (Table 1). More than 50% of β-glucosidase was present in the cell debris fractions.

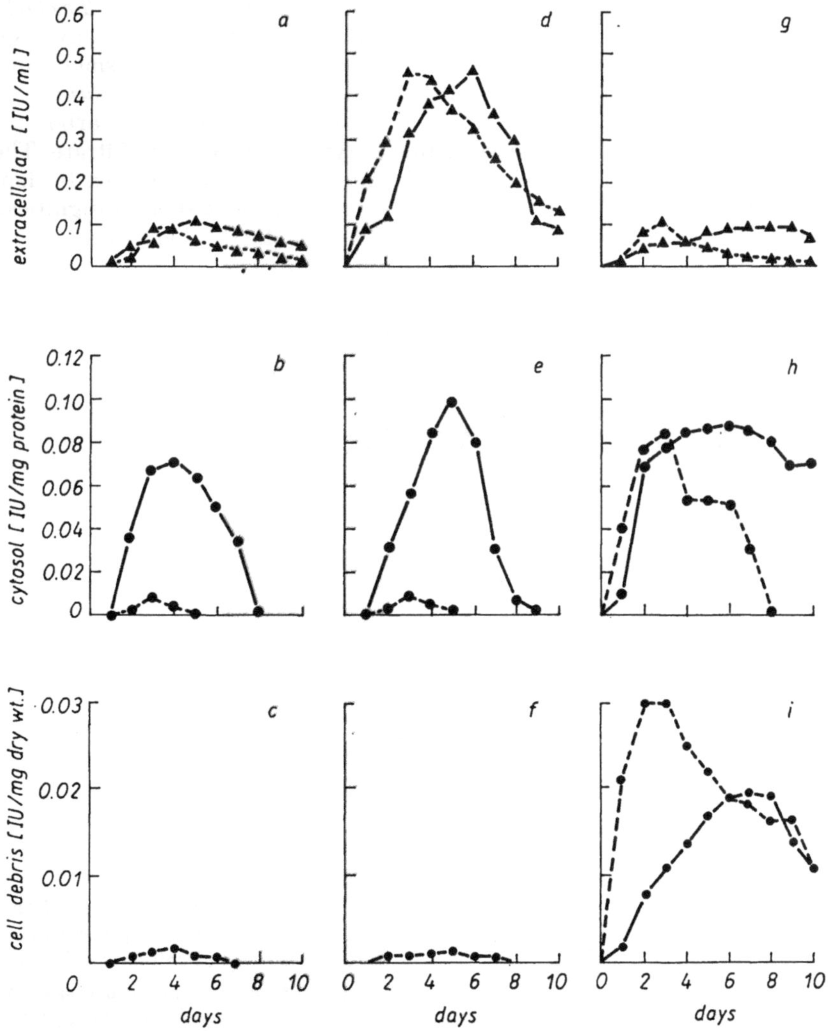

Fig. 1 Production of FPactivity (a—c), CMcellulase (d—f), β-glucosidase (g—i) in shake (---) and stationary (——) cultures on 1% carboxymethyl cellulose.

Table 1. Distribution of enzymes in the three fractions of cellulases

Enzyme fraction	Enzyme distribution [units per flask]		
	FPactivity	CMcellulose	β-glucosidase
Extracellular	2.75	10.25	2.12
Cytosol	0.67	0.096	0.095
Cell debris	0.27	0.39	2.83

In another study on the two strains of *T. pseudokoningii* grown on lactose, β-glucosidase was mainly found in the cell bound state [14]. HALLIWELL and LOVELADY [15] have reported that 99% of CMcellulase in *T. koningii* is extracellular, whereas 90% of β-glucosidase is associated with cells. Studies on fungi other than *Trichoderma* which include *Penicillium ianthinellum* and *Sporotrichum pulverulentum* have revealed that β-glucosidase and β-glucanases are predominantly cell free with small amounts being associated with the mycelium [16, 17]. In the case of *S. pulverulentum* only cell wall bound β-glucosidase was formed when cellobiose was used as the carbon source, while cellulose induced increasing release of this enzyme in the culture filtrate. Therefore, the release of enzyme or its association with mycelium has been shown to depend on the carbon source [18—20]. The cytosol enzymes represented an insignificant fraction except for FPactivity which contributed 18.5% of the total enzyme.

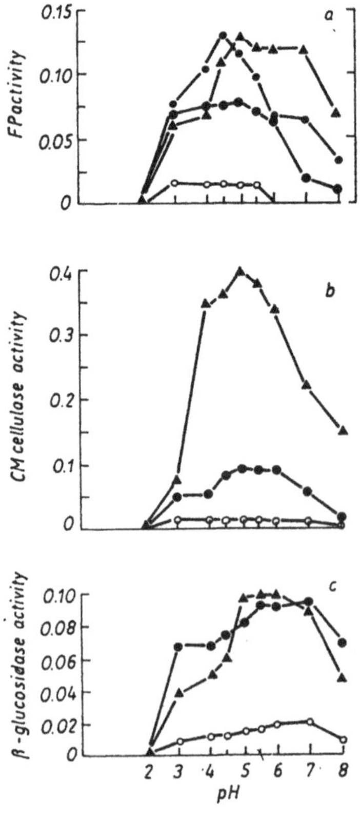

Fig. 2. Effect of initial pH of the medium on growth and cellulase production in extracellular, cytosol and cell-debris fractions.
▲— extracellular [IU/ml]; ●— cytosol [IU/mg protein]; ○— cell-debris [IU/mg dry wt. x 10^{-1}]; •— dry wt. [mg/flask].

When the initial pH of the medium was adjusted between 2—6 a continuous increase in mycelial dry weight along with three enzyme components was observed in all the three fractions. The optimal pH for extracellular FPactivity and CMcellulase was 5.0 and for β-glucosidase 6.0 (Fig. 2). The fungus could produce fair amounts of FPactivity and β-glucosidase in the extracellular fraction even at pH 7.0 whereas CMcellulase was reduced to half at this pH showing its sensitivity to higher pH values. The yields of the cytosolic and cell debris enzymes were maximum between pH 3—7 (Fig. 2) and persisted even at 8.0 except for FPactivity in the cell debris fraction which disappeared at pH 6.0.

The production of extracellular enzymes and mycelial growth was maximum at 27 °C (Fig. 3). The optima for enzymes in the cytosol and cell debris fraction were observed at 22 °C and 27 °C and were completely absent at higher temperature except for β-glucosidase which was prevalent at all temperatures excepting 42 °C (Fig. 3c). The optimal pH and temperature recorded for enzyme production are in accord with earlier observations [3, 7, 21].

One of us (MKK) is thankful to the Council of Scientific and Industrial Research, New Delhi for the grant of senior research fellowship.

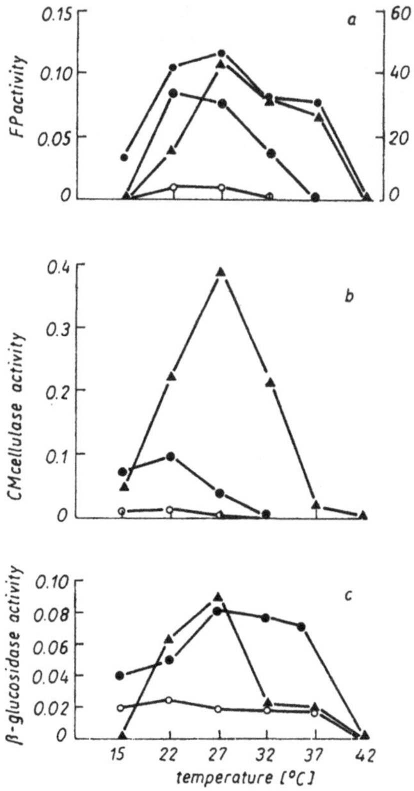

Fig. 3. Effect of temperature of incubation on growth and cellulase production in extracellular, cytosol and cell-debris fraction.

▲— extracellular [IU/ml]; ●— cytosol [IU/mg protein]; ○— cell-debris [IU/mg dry wt. × 10^{-1}]; •— dry wt. [mg/flask].

References

[1] Mandels, M., Weber, J.: Adv. Chem. Ser. **95** (1969), 391.
[2] Sternberg, D.: Biotechnol. Bioeng. Symp. **6** (1976), 35.
[3] Taniguchi, M., Tanaka, M., Matsuno, R., Kamikuizo, T.: J. Ferment. Technol. **58** (1980), 43.
[4] Bisaria, V. S., Ghose, T. K.: Enzyme Microb. Technol. **3** (1981), 90.
[5] Reese, E. T.: Biotechnol. Bioeng. Symp. **5** (1975), 77.
[6] Mandels, M., Andreotti, R. E.: Process Biochem. **13** (1978), 6.
[7] Mukhopadhyay, S. N., Malik, R. K.: Biotechnol. Bioeng. **22** (1980), 2237.
[8] Sandhu, D. K., Sidhu, M. S., Sandhu, R. S., Kalra, M. K. — In: Mukerji, K. G., Agnihotri, V. P., Singh, R. P. (Eds.), Progress in Microbial Ecology. 1984, p. 525.
[9] Sandhu, D. K., Kalra, M. K.: Trans. Br. Mycol. Soc. **79** (1982), 409.
[10] Updegraff, D. M.: Anal. Biochem. **32** (1969), 420.
[11] Lowry, O. H., Rosenbrough, N. J., Farr, A. L., Randall, R. J.: J. Biol. Chem. **193** (1951), 265.
[12] Boretti, G., Garofano, L., Montecucchi, P., Spala, G.: Arch. Microbiol. **92** (1973), 189.
[13] D'Souza, J., Furtado, I.: Indian J. Microbiol. **17** (1977), 194.
[14] Kubicek, C. P.: Canad. J. Microbiol. **29** (1983), 163.
[15] Halliwell, G., Lovelady, J.: J. Gen. Microbiol. **126** (1981), 211.
[16] Eriksson, K.-E., Hamp, S.: Eur. J. Biochem. **90** (1978), 183.
[17] Rapp, P., Grote, E., Wagner, F.: Appl. Environ. Microbiol. **41** (1981), 857.
[18] Berg, B.: Canad. J. Microbiol. **21** (1975), 51.
[19] Berg, B., Pettersson, G.: J. Appl. Bacteriol. **42** (1977), 65.
[20] Chaudhary, K., Tauro, P.: J. Appl. Microbiol. Biotechnol. **15** (1982), 185.
[21] D'Souza, J., Volfova, O.: Eur. J. Appl. Microbiol. Biotechnol. **16** (1982), 123.

Acta Biotechnol. 6 (1986) 2, 167–174

Production, Characterization and Use of Expendable Kefir Starter Culture in the Kefir Production Process

KRAMKOWSKA, A., FESNAK, D., KORNACKI, K., BAUMAN, B.

Central Administration of Dairy Industry
Production Laboratory of Dairy Biopreparations
ul. Warszawska 111, 10-957 Olsztyn, P.O.B. 1, Poland

Summary

Upon sterilization, freeze-dried kefir grains are to be supplemented with yeast preparation, yeast showing the lowest survival in the course of freeze drying.
A mixture of 20% sucrose solution and starch appeared to be the most efficient dispersing agent to protect yeast in the freeze-drying process.
A starter which is produced with the kefir starter culture, does not differ in its microbiological, biochemical and organoleptic properties from a starter as obtained with kefir grains in the same time, such starter shows a better consistency.
The preparation of such starters is less laborious and less complicated than the preparation of starters using kefir grains.

Introduction

The interest for fermented milks is connected with the theory proposed by METCHNIKOV that the growth of the putrefactive bacteria is inhibited in the intestines by lactic acid bacteria. The recent findings of science have confirmed that those products are characterized by valuable dietetic properties which are reflected in their chemical, microbiological and vitamin composition, favourable for the human organism. The most utilized fermented milks are yoghurt and kefir [16, 17].
Kefir is obtained by the complexe lactic and alcoholic fermentation in milk. In its production, kefir grains are involved, which have the appearance of rice grains cooked to rags, of irregular shape, elastic to the touch [5, 8, 16].
The kefir grain microflora is no uniform culture, but it may contain various kinds of microorganisms, according to the cultural conditions. In general, three groups of microorganisms are found to occur: streptococci, bacilli and yeasts. They involve mesophilic acidifying bacteria, *Str. lactis*, *Str. cremoris*; aromaforming bacteria, *Str. diacetilactis* and *Leuconostoc*; *Lactobacilli*, *Lbc. caucasicus* and acetic bacteria [9]. Of yeast, spore-forming species are present, such as *Saccharomyces*, *Schizosaccharomyces*, which utilize lactose directly, as well as non-spore forming and non-lactose-fermenting yeasts of *Torula* species [8, 16].
In kefir production, the greatest difficulties are provided by the kefir grain cultivation. To obtain a product of proper quality, the following condition are to be observed: optimum temperature of grain cultivation must be maintained; milk must be changed

in grains every day, to maintain milk-grain ratio 1 : 15 to 1 : 20; overacidity is to be avoided and extreme caution is to be applied to prevent infection. The studies on troubles of kefir grain cultivation permitted the cases of their occurance to be partially determined. A cause of a strong gas formation was found to be the overgrowth of *Leuconostoc dextranicum* bacteria and yeasts. The presence of foreign yeast strains may result in fruit flavour and an excessive growth of acetic bacteria may lead to slime in grains. Slimy grains lose their characteristic elasticity and a mealy bloom is formed on their surface. That defect may be caused by the mould *Geotrichum candidum*. Grains are deacidified by the mould growing on them and they become suspectible to the action of putrefactive bacteria. On industrial conditions, the Coli bacteria infection may be caused by ineffective pasteurization of starter milk or by reinfection, the growth of those bacteria cause an unclear flavour and excessive gas formation. Such troubles as poor acidification or whey secretion are often caused by chemical and physical changes in milk [1].

The observation of all principles for kefir grain cultivation requires diligence, some length of time, appropriate laboratory conditions and a working staff properly skilled.

In addition to the above inconveniences, the use of kefir grains appear to be troublesome because of the short time of expiration. In viable form, the kefir grains are active when just produced and they conserve their keeping quality for 5 days at ambient temperature. Unfavourable transport conditions or exceeded expiration time (5 days) cause the kefir grains to lose their activity [7].

The above inconveniences in kefir grains cultivation impelled the studies to be undertaken, intending to produce a starter culture which would eliminate the kefir grains cultivation in dairy plants. These attempts resulted in a freeze-dried kefir preparation being obtained which would replace the kefir grain starter, as prepared by traditional procedure.

The first satisfactory results enabled the continuation of studies on the freeze drying possibility of kefir starter cultures [10]. The greatest difficulties were provided by a low survival of yeasts in the course of stabilization [4]. Hence it appeared to be necessary for yeasts, to be propagated and stabilized separately and then to be added in such form, to the freeze-dried kefir grains in which a high percentage of yeast had been removed.

Experimental and Methods

Preparation of Kefir Grains for Stabilization

The principle of kefir grains preparation consists in combining the duly prepared kefir with viable kefir grains. Kefir was obtained from sterile milk, inoculated with 3% of kefir starter, as produced from a kefir grain culture, and then incubated at 23 °C for 20 hours. Twice during incubation, the kefir was neutralized with sodium hydroxide up to the acidity of 8—10 °SH. Viable kefir grains were mixed with kefir and then 18% lactose solution was added. When preparing the mixture, the following ratio of grains, starter and lactose was observed: 1 : 1 : 0.5. The final acidity of ready-made preparation was 16—20 °SH.

Preparation of Kefir Yeasts

A superficial cultivation of yeasts, as isolated from kefir grains was run on a wort-agar medium at 23 °C for 48—72 hours. Superficially grown yeasts were rinsed with sterile 20% sucrose solution and mixed with sterilized starch.

Control of Preparations before Stabilization

In the kefir grain mixture, quantitative percentage of the following groups of microorganisms was determined:

- acidifying mesophile streptococci, on the medium according DEMETER [8],
- aroma-forming mesophile streptococci, on the NICKELS-LEESMENT (N-L) medium [14],
- lactobacilli, on MRS medium [3],
- yeasts, on the wort-agar medium [8].

In kefir yeasts preparation, counts of yeast germs in 1 ml were determined [8].

Stabilization of Preparations Obtained

Kefir starter culture obtained was stabilized using two procedures, freezing and freeze-drying:

- freezing of the preparation in an alcohol bath, type CD-165, of the Heto Company (Denmark) at temperature —50 °C, and then hardening at —40 °C for 6—8 hours,
- freeze-drying — frozen preparations were freeze-dried using a Type CD-155 freeze-dryer, produced by the Heto Company (Denmark), in vacuum of the order of 10^{-2} for about 24 hours. Final drying temperature did not exceed 40 °C.

Control of Dry Preparation

Dried kefir grain and yeast preparations were desintegrated in a grinder and then the following determinations were made:

- microbiological purity — presence of foreign microflora, including Coli-bacteria [8],
- presence of various groups of microorganisms as under „Control of Preparations before Stabilization",
- milk coagulation time, when inoculated with kefir grain preparation at 23 °C,
- organoleptic properties of milk, pasteurized at 95 °C/30 min. inoculated with the kefir grain preparation and incubated at 23 °C for 20—24 hours.

Standardization of Ready-Made Product

For producing kefir starter culture, to be used for inoculation of 3 l milk of kefir (mother) starter:

300 mg of kefir grain preparation,
300—600 mg of yeast preparation,
300—600 mg of complement

were used, depending on counts of microorganisms.

Quality and Keeping Quality Testing of Freeze Dried Kefir Starter Culture

The quality of starter cultures was checked twice during storage. Counts of microorganisms were tested on media, as under „Control of Preparations before Stabilization". Simultaneously, milk was inoculated with a freeze-dried starter culture, previously pasteurized at 95 °C for 30 min. and incubated at 23 °C for 20—24 hours. Parallelly, milk was inoculated with a traditional kefir starter. In starters obtained, fermentation and biochemical properties were compared. The assays involved:

- milk coagulation time at 23 °C,
- titrable and active acidity,
- fermentation activity, using TTT-titrator of the Radiometer Company (Denmark), pH changes of milk being tape-recorded during 24 hours incubation at 23 °C,

- organoleptic examination,
- acetyl producing ability, according to the PIEN's method [15],
- joint acetoin and diacetyl producing ability, by the method of BRANDL [2],
- protein degrading ability, up to N-alpha-amino acids, by the method of LEE and TAKAHASHI [12],
- volatile fatty acid content (VFT), by the method of HEMPENIUS and LISKA [6],
- acetaldehyde production, by the method of LINDSAY and DAY [13],
- ethanol and acetone production, by gas chromatography, using a gas chromatograph of Pay-Unicam Company [England).

Results and Discussion

The studies were intended to obtain an expendable kefir starter culture which could be substituted for the traditional kefir grain cultivation, and then to compare its properties with those of kefir grains. The manufacturing conditions were chosen in that manner, that, on vivifying a kefir starter preparation, a starter could be obtained, comparable with kefir starter, as produced from the traditional kefir grain cultivation. To this end, the quantitative composition of individual groups of microorganisms was determined in the preparation before and after freeze-drying, as well as in starters obtained from the dried preparation. Based on the results of these assays, the survival of acidifying bacteria, aroma forming bacteria, lactobacilli and yeasts was found to be 50%, 44.7%, 16% and 5%, respectively (Table 1).

Table 1. Survival of the microflora in the kefir grain preparation and yeast preparation in stabilization process

Stabilization procedure	Acidifying bacteria germs in 1 ml	Survival [%]	Aroma forming bacteria germs in 1 ml	Survival [%]	Lacto-bacilli germs in 1 ml	Survival [%]	Yeast germs in 1 ml	Survival [%]
Liquid	4.6×10^7	—	2.9×10^7	—	6.9×10^7	—	8.1×10^4	—
Frozen	3.3×10^7	73.9	2.3×10^7	79.3	1.2×10^7	17.4	1.2×10^4	14.8
Freeze-dried	2.3×10^7	50.0	1.3×10^7	44.7	1.1×10^7	16.0	4.2×10^3	5.1

Good survival results were obtained for mesophile bacteria strains in the course of freezing process. The survival of acidifying bacteria and aroma forming bacteria was 73.9% and 79.3%, respectively. The highest losses during freezing were observed for yeasts (85.2%) and *lactobacilli* (82.6%). A low survival of yeasts during stabilization is probably caused by the fact that yeast germs are larger than those of bacteria, and therefore they are more exposed to low temperature in the course of freezing. Ice crystals are formed which destroy cell walls of yeasts [4].
In this connection, it appeared to be necessary, to supplement the kefir grain preparation with yeasts, in the amount which would be sufficient to compensate losses occurring in the course of stabilization. Satisfactory results were obtained in the course of freezing and storage, when rinsing the surface growth of yeasts on wort-agar with 20% sucrose solution. The yeast rinsings were combined with sterilized starch to obtain the consistency of thick pulp. Kefir grain and yeast preparations were produced in triplicate, to prepare 3 variants of kefir starter cultures. The starter cultures were stored at +5 °C for a period of 3 months. Results of assays, just produced and after 6 weeks of storage, are presented in Table 2.

Table 2. Changes of microbial counts in the kefir starter culture during 6 weeks of storage at +5 °C

Starter cultures (variants)	Acidifying bacteria germs in 1 g of d.m.		Aroma forming bacteria germs in 1 g of d. m.		Lactobacilli germs in 1 g of d. m.		Yeasts germs in 1 g of d. m.	
	a	b	a	b	a	b	a	b
I	3.2×10^7	1.3×10^7	1.0×10^7	6.0×10^6	2.5×10^8	1.5×10^8	9.8×10^4	4.8×10^4
II	3.8×10^7	7.2×10^6	4.0×10^6	2.9×10^6	7.5×10^8	7.9×10^7	3.6×10^5	9.6×10^4
III	2.6×10^7	9.0×10^6	5.2×10^6	2.5×10^6	6.4×10^7	1.5×10^7	2.5×10^4	4.0×10^3

a — Microbial counts after production
b — Microbial counts after 6 weeks storage

As it appears from the counts obtained, the average germ counts of organisms decreased, in all groups, by one power during storage, but they were still comparable to the respective microbial counts, as present in the microflora of viable kefir grains.
In the same time, biochemical and fermentation properties as well as organoleptic characteristics of kefir starter cultures were estimated and compared with those of the control starter (Table 3).

Table 3. Estimation of biochemical properties of starters as produced with starter cultures

Starter culture (variants)	pH		Acetoin [mg/l]		Diacetyl [mg/l]		Amino acid nitrogen [mg/l]		Acet-aldehyde [mg/l]		VFA — ml 0.1n NaOH/100 ml sample	
	a	b	a	b	a	b	a	b	a	b	a	b
I	4.71	4.53	8.96	244.4	1.74	0.595	43.26	42.5	0.34	0.21	7.8	7.8
II	4.73	4.44	18.20	259.4	0.95	0.863	32.05	44.2	0.39	0.29	9.4	9.4
III	4.72	4.45	16.10	245.6	1.23	0.943	46.29	41.8	0.28	0.26	9.6	8.8
IV*	4.41	4.35	38.5	211.0	0.56	0.949	63.49	45.7	0.33	0.34	8.6	8.4

a — just produced
b — after 6 weeks of storage
* — control starter, as produced with a traditional kefir grain culture

The analysis of biochemical properties, made in duplicate, showed only slight differences between starters, as produced from starter cultures, and those prepared by traditional procedure.
The presence of volatile fatty acids (VFA) is the most apparent of biochemical characteristics of kefir studies [11]. When just prepared, the VFA content of starter cultures ranged between 7.8—9.6 ml 0.1 n NaOH/100 ml in test samples and 8.6 ml 0.1 n NaOH in a control sample. After storage, the respective amounts were 7.8—9.4 ml 0.1 n NaOH/100 ml and 8.4 ml 0.1 n NaOH/100 of sample.
The test samples were characterized by producing aroma substances at the middle level. Upon producing, the acetoin content of starter cultures was low and it ranged from 8.96 to 18.2 mg/l in test samples, as compared to 38.5 mg/l in control sample. After storage, those amounts increased to 244.4—259.4 mg/l and 211 mg/l, respectively. The diacetyl content was higher upon production and it was at the level of 0.95 to 1.74 mg/l in test samples and 0.56 mg/l in a control sample. After storage, the amount

of that compound decreased to 0.595—0.943 mg/l in test samples, and it was 0.949 mg/l in the control sample. These changes may be only apparent, because assays were carried out with fresh bulk milk, the chemical composition of which varies depending on the way of feeding and the season [9]. It may be stated that both after production and after storage, there are no essential differences between test samples and the control sample in producing aroma substances.

In samples assayed, a low but even content of acetaldehyde was found, at the level of 0.28—0.34 mg/l, and 0.21—0.34 mg/l after storage. That amount provides a proper test of kefir.

The kefir starter culture may be used in a frozen or freezedried form. In view of more convenient storage and lower transport costs, freeze-dried starter cultures are exclusively used in the dairy industry.

Biochemical properties of starters and kefir, as produced with freeze-dried and frozen starter cultures are compared in Table 4.

Table 4. Characterization of biochemical properties of kefir starter cultures and kefir as produced with them

Starter culture (variant)	pH	Acetoin [mg/l]	Diacetyl [mg/l]	VFA — ml 0.1 n NaOH/100 ml sample	Ethanol [mg/l]	Acetone [mg/l]
I — freeze-dried	4.44	259.0	0.863	9.4	17.5	6.5
II* — freeze-dried	4.30	310.0	0.675	9.8	11.8	3.7
II — frozen	4.49	254.0	0.280	11.8	18.8	4.3
II* — frozen	4.31	266.0	0.650	11.6	23.8	3.3
III — control	4.35	211.0	0.949	7.6	45.2	7.4
III* — control	4.33	245.0	0.836	14.2	17.9	3.4

* — kefir produced on the first starter — kefir

The alcoholic fermentation is a characteristic property of kefir grain microflora. When obtained with a freeze-dried starter culture, the kefir starter was characterized by the ethanol content of 0.17%, while the ready-made kefir contained 0.11% of that compound. When obtained with a frozen starter culture, the starter and kefir contained 0.18% and 0.23% of ethanol, respectively.

The highest ethanol content was shown in the starter as produced with the control sample — 0.45%, while the kefir produced by means of that starter contained 0.17% ethanol. These results were in agreement with the findings of KOROVKINA [11] who reported, following other authors, that the ethanol content of one day kefir was 0.12%. She informs, too, that the best organoleptic characteristics were found in kefir and kefir starter which contained 0.16—0.21% and 0.18—0.40% ethanol, respectively.

In these studies, kefir starters were obtained in which free fatty acid contents were 9.4 and 11.8 ml 0.1 n NaOH/100 ml of a sample, with dry and frozen preparation, respectively, as compared with 7.6 ml 0.1 n NaOH/100 ml for a control sample. For kefir, as obtained with those starters, the respective amounts were: 9.8, 11.6, and 14.2 ml 0.1 n NaOH/100 ml of sample, respectively. The product studies showed diacetyl and acetoin content which ranged between 0.280—0.949 ml/l and 211.04—310 mg/l, respectively.

The results of these studies showed, that the kefir starter culture, both in frozen and freeze-dried form, enables to obtain a kefir which does not differ, by its test properties, biochemical characteristics and composition of individual groups of microflora, from the product, as obtained by traditional method (Table 5).

Table 5. Characterization of fermentation properties and organoleptic features of kefir

Starter culture (variant)		Coagulation time [h]	VP test	Organoleptic characteristics	Microscopic picture
I — freeze-dried	4.90	22	+	Moderately acid, clean, cream consistency, effervescent, sl. refreshing, flavour of kefir grains	Short chains of streptococci, 2—3 germs of yeast per several fields of view, few lactobacilli
I* — freeze-dried	4.78	16	+	Acid, clean, cream consistency, effervescent, sl. refreshing, aromatic flavour of kefir grains	
II — frozen	4.49	16	+	Moderately acid, cream consistency, sl. refreshing, effervescent, flavour of kefir grains	Short chains of streptococci, diplococci, 2—3 germs of yeast per several fields of view, 1—2 lactobacilli per several fields of view
II* — frozen	4.31	16	+	Acid, clean, cream consistency, effervescent, sl. refreshing flavour of kefir grains	
III — control	3.68	20	+	Acid, thick cream consistency, effervescent, refreshing, flavour of kefir grains	Streptococci in form of diplococci or short chains, 2—3 germs of yeast per several fields of view
III* — control	3.90	16		Acid, clean, regular consistency, less viscid, sl. effervescent, flavour of kefir grains	

* — kefir produced on the first starter — kefir

On commercial conditions, the use of kefir starter in frozen form would be possible, provided that convenient containers would be available for production and low temperature could be assured during transport and storage. Starter cultures in dry form do not require any special caution or protection, no changes occurring in their propertie, transport or storage.

When produced with the expendable starter culture, the kefir shows a great reproducibility of properties. It is much more appreciated for its homogeneous, viscid consistency, no tendency for delamination and more pronouhced, refreshing taste. When following instructions, the starter culture ensures an active bulk starter to be obtained in commercial conditions.

Because of being less laborious and less complicated, the bulk starter preparation using the expendable starter culture is decidedly preferable to the traditional starter, and those advantages speak for using those starter culture for kefir production.

The Use of Starter Cultures

From one portion of starter culture, 3 litres of mother starter may beprepared. That starter may be used to inoculate milk for kefir preparation, or to prepare a bulk starter in dairy plants in which larger amounts of kefir are produced.
To inoculate milk for making kefir or bulk starter, 3% of inoculum is used.

Received May 3, 1984

References

[1] BAWINA, N. A., ROZKOWA, J.: Mołocznaja Prom. **4** (1974), 18.
[2] BRANDL, E.: Milchwissenschaft Ber. **4** (1960), 289.
[3] BURBIANKA, M., PLISZKA, A.: Mikrobiologia żywności (1971).
[4] GÓRZYŃSKA, J.: Przemysł fermentacyjny i rolny. **11** (1971) 7.
[5] HATTOWSKA, H.: Mikrobiologia mleczarstwa (1964).
[6] HEMPENIUS, W., LISKA, B.: J. Dairy Sci. **51** (1966), 221.
[7] Instrukcja stosowania szczepionki jednorazowego użycia (1980).
[8] KARNICKA, H., KNAUT, T.: Mikrobiologia ogólna (1964).
[9] KAROLEWA, N. S.: Techniczeskaja mikrobiołogia celnomołocznych produktow. **126** (1975).
[10] KORNACKI, K., KRAMKOWSKA, A., FESNAK, D., BAUMAN, B.: Przegląd mleczarski. **4** (1980),13.
[11] KOROWKINA, Ł. N., PATKUL, T. M.: Mołocznaja Prom. **6** (1976), 18.
[12] LEE, P., TAKAHASHI: Annal. Biochem. **14** (1966), 71.
[13] LINDSAY, B., DAY, A.: J. Dairy Sci. **48** (1965), 665.
[14] NICKELS, W. E., LEESMENT, H.: Milchwissenschaft. **19** (1964), 374.
[15] PIEN, J.: Lait. **27** (1975), 119.
[16] PIJANOWSKI, E.: Zarys chemii i technologii mleczarstwa (1980).
[17] SMETAŃSKI, R.: Sfermentowane napoje mleczne (1948).

Acta Biotechnol. 6 (1986) 2, 175—187

The Effect of Automatic pH Maintenance on Steady State Cultivation Regimes and Their Stability
A Theoretical Study of the Case of a Perfect pH Controller and Perfect Stirring

MINKEVICH, I. G.

U.S.S.R. Academy of Sciences
Institute of Biochemistry and Physiology of Microorganisms
142292 Pushchino, Moscow Region, U.S.S.R.

Summary

The dynamics of pH-controlled chemostat and pH-auxostat is considered. The medium flow governed by pH-controller is calculated using the ionic (mass-charge) balance equations. The generalized equations of pH-controlled continuous fermentation are developed for the case of two variables: cell biomass and residual substrate concentrations. The study of steady state solutions has shown a substantial dependence of the variable patterns on a concentration of alkali (or acid) in titrant medium flow. The generalized analytical expressions for eigenvalues (reciprocals of transient characteristic times) of the culture dynamics have been obtained. The rates of transients during the steady state establishment are considered using these eigenvalues.

Introduction

In most cases, continuous cultivation of microorganisms is employed to provide steady state growth of microbial culture. The steady state stability includes two aspects:

i) a possibility of operation during sufficiently long time (theoretically, infinitely long), and
ii) the rate of achievement of the steady state after start-up or a deviation.

The second point is important for actual constancy of the steady state because a real equipment for cultivation, especially a large-scale one, possesses finite precision of pumps and controllers and imperfect (to a different extent) stirring.
The simplest cases of dynamics of a continuous culture take place in chemostat limited by a single substrate and in turbidostat. The dynamics of such chemostat was studied theoretically by many authors [1—7]. The analytical study of the chemostat stability was reported in [1, 2, 5, 6], where LYAPUNOV's method was applied and the roots of a characteristic equation were found as functions of dilution rate, fed substrate concentration and physiological parameters of culture. In [6] it was shown that the characteristic values of the process of chemostat steady-state approach usually had highly different numerical values which display themselves as time scaling of transients. We are not familiar with any quantitative investigations of turbidostat stability.
In the mentioned analyses it was usually accepted that the fed medium was supplied through a single inlet. However, a real fermenter operating in chemostat regime, as a

rule, has another inlet through which a titrant flow (base or, if necessary, acid) is supplied for maintenance of a preset pH-value. While the nutrient flow is set constant at a desirable value, the titrant flow is established by an automatic controller. The total flow value is the sum of the both, which should be taken into account when calculating the dilution rate. The share of the titrant flow in the total one depends on the base (or acid) concentration and increases as the latter decreases. The presence of the titrant flow was not taken into account in earlier studies of chemostat stability.

Another continuous cultivation regime, at which the value of inflow is related to pH control, is pH-auxostat described first by PECHURKIN et al. (as „*pH*-stat") [8] and later by MARTIN and HEMPFLING [9]. In *pH*-auxostat the nutrient flow is simultaneously the titrant flow. Therefore, the complete inflow is supplied by means of a *pH*-controller. To prevent precipitation of salts insoluble at high *pH*, the medium should be divided into two or more parts supplied through separate inlets. However, the corresponding pumps are controlled by one and the same controller and operate synchronously in a fixed proportion. Therefore, kinetically these several flows proceed as a unite flow.

pH-Auxostat is employed for cultivation at maximal specific growth rate or in the region of substrate inhibition, i. e. under conditions when chemostat cannot run. *pH*-Auxostat is an analogue of turbidostat, since the stability of both cannot be provided by cells and essentially maintained by the corresponding controller. On the other hand, as it follows from experience, *pH*-auxostat maintains the value of dilution rate much better than turbidostat, since *pH* is a low-inertial characteristic compared with cell biomass concentration.

This work is devoted to mathematical analysis of the stability of cell biomass and the residual substrate concentrations in *pH*-stated chemostat and *pH*-auxostat. The effect of a titrant concentration on the steady-state values of these concentrations as well as on the other cultivation variables is also considered.

Influence of ion balance of the culture upon the rate of medium supply

This effect is connected with *pH* control. We consider here the case of perfect (ideal) *pH* control providing the exact maintenance of a constant preset *pH* value in fermenter. Perfect stirring in fermenter bulk is also supposed.

In the previous work [10, 11] we derived the equations of ion balance (see below) from the fact of high-rate establishment of ionic equilibria as well as mass and charge balances including the condition of electroneutrality.

Denotations

l, number of species of a substance, independent of its ionic state (e.g. phosphoric acid as well as all forms of phosphate ions have one and the same l number; $l = 1$ is H^+, $l = 2$ are OH^- and H_2O, etc.); below we shall use sometimes subscripts instead of the corresponding numbers;

C_l, mol · l^{-1}, the lth substance concentration in fermenter; $C_l(1)$ and $C_l(2)$ are concentrations in media supplied via inlets 1 and 2. $C_1 = [H^+] = 10^{-pH}$, proton concentration and corresponding *pH* value in fermenter; $C_1(1) = 10^{-pH(1)}$ and $C_1(2) = 10^{-pH(2)}$ are proton concentrations and *pH* values in media 1 and 2;

X, $g \cdot l^{-1}$, dry cell biomass (DCB) concentration in fermenter;

$T_l(C_1)$, the average charge of the lth substance molecules (ions) in fermenter at the given C_1 (i.e. *pH*) value, equivalents of proton charge units (PCU) per mol (e.g. at $pH \sim 4 \div 5$ the total phosphate is $H_2PO_3^-$, and $T_{phosphate} = -1$); $T_l(C_1(1))$ and $T_l(C_1(2))$ are analogous quantities for inlet media 1 and 2 at $pH(1)$ and $pH(2)$;

q, *eq.* PCU · (g DCB)$^{-1}$, the specific charge of viable cell biomass per cell dry matter (depends on *pH*);

λ_l, mol · (g DCB)$^{-1}$, stoichiometric coefficient of the lth substance uptake ($\lambda_l > 0$) or excretion ($\lambda_l < 0$) due to cell growth;

δ_l, mol · (g DCB)$^{-1}$ · h^{-1}, the rate of the lth substance uptake ($\delta_l > 0$) or excretion ($\delta_l < 0$) due to cell maintenance and futile dissipation processes [12];

λ_q, *eq.* PCU · (g DCB)$^{-1}$, stoichiometric coefficient of formation of cell charge during cell biomass growth;

(it is appropriate to mention here that in general case the dynamics of cell charge depends also on δ_q, a quantity analogous to δ_l [10,11], but under usual growth conditions the process of cell maintenance and futile dissipation does not lead to charge accumulation or disappearance, i.e. $\delta_q = 0$);

μ, h^{-1}, specific rate of cell biomass growth;

D, h^{-1}, total dilution rate;

$D(1)$ and $D(2)$, h^{-1}, components of D, connected with media flows via inlets 1 and 2 ($D = D(1) + D(2)$);

F_l^0, mol · l^{-1} · h^{-1}, rate of supply ($F_l^0 > 0$) or evaporation ($F_l^0 < 0$) of electroneutral form of lth substance due to aeration.

The equations of mass-charge balance are [10, 11]:

$$C_1 - \frac{K_W}{C_1} + \sum_{l \geqq 3} T_l(C_1)\, C_l + qX = 0 \tag{1}$$

(electroneutrality condition; $K_W = 10^{-14}$ is the constant of water dissociation),

$$\frac{dC_l}{dt} + DC_l = -\lambda_l \mu X - \delta_l X + D(1)\, C_l(1) + D(2)\, C_l(2) + F_l^\circ \tag{2}$$

(balance of lth substance, $l \geqq 3$),

$$\frac{dq}{dt} = -\mu q + \lambda_q \mu \tag{3}$$

(cell charge dynamics). The known equation of cell biomass kinetics is also needed:

$$\frac{dX}{dt} = \mu X - DX \tag{4}$$

Here and below the working volume of fermenter is assumed constant. Denoting the left-hand side of Eq. (1) as A and utilizing $C_1 =$ Constant (due to automatic pH control),

$$\frac{dA}{dt} + DA = D\left(C_1 - \frac{K_W}{C_1}\right) + \sum_{l \geqq 3} T_l(C_1) \left(\frac{dC_l}{dt} + DC_l\right)$$
$$+ q\frac{dX}{dt} + X\frac{dq}{dt} + DqX = 0 \tag{5}$$

From Eqs. (3) and (4):

$$q\frac{dX}{dt} + X\frac{dq}{dt} + DqX = \lambda_q \mu X \tag{6}$$

Substituting Eqs. (2) and (6) into Eq. (5),

$$D(1)\left[C_1 - \frac{K_W}{C_1} + \sum_{l_{sub}} C_l(1)\,T_l(C_1)\right] + D(2)\left[C_1 - \frac{K_W}{C_1} + \sum_{l_{sub}} C_l(2)\,T_l(C_1)\right]$$
$$- \mu X \sum_{l_{sub}} \left(\lambda_l + \frac{\delta_l}{\mu}\right) T_l(C_1) - \mu X \sum_{l_{pr}} \left(\lambda_l + \frac{\delta_l}{\mu}\right) T_l(C_1)$$
$$+ \mu X \lambda_q + \sum_l F_l{}^0 T_l(C_1) = 0 \tag{7}$$

where l_{sub} and l_{pr} are numbers of substrates and products, correspondingly.
The third and the subsequent terms in Eq. (7) are connected with uptake of charged substrates, formation of charged products, formation of the electric charge of cells and with evaporation of neutral forms of partially ionized substances.
Electroneutrality of the supplied media leads to

$$C_1(1) - \frac{K_W}{C_1(1)} + \sum_{l_{sub}} C_l(1)\,T_l(C_1(1)) = 0 \tag{8}$$

$$C_1(2) - \frac{K_W}{C_1(2)} + \sum_{l_{sub}} C_l(2)\,T_l(C_1(2)) = 0 \tag{9}$$

From Eqs. (8) and (9) it can be seen that the first or second term in Eq. (7) equal to zero if $pH(1)$ or $pH(2)$ in the media equals to pH in fermenter. Simultaneous equality of both terms to zero ($pH(1) = pH(2) = pH$) is a rare case when growth ($\mu X > 0$) can be maintained due to partial evaporation of acidic substrate. This theoretical case is of no interest.
Let $D(2)$ be the inflow of titrant per liter of the working volume of fermenter. Its value can be easily calculated from Eq.(7). Here we shall consider the case when: 1) $pH(1)$ in the medium supplied with the set rate via inlet 1 is equal to pH in fermenter; then the coefficient at $D(1)$ is zero; 2) evaporation of partially ionized organic substances is absent, i. e. the corresponding $F_l{}^0 = 0$; and 3) pH in fermenter is acidic or somewhat below neutral, so that carbon dioxide accumulation in broth is small and the total carbon dioxide formed is removed by air flow [13]; this means:

$$-\mu X \left(\lambda_{CO_2} + \frac{\delta_{CO_2}}{\mu}\right) + F^0_{CO_2} = 0$$

and carbon dioxide can be excluded from Eq. (7). Within this pH region evaporation of ammonia is absent as well. The remaining inorganic substances are not volatile.
The situation under the above assumptions can be widely met in practice. If this is the case, Eq. (7) is essentially simplified compared with the general form.
The coefficient at D(2) can be rearranged. Let the medium of inlet 2 be titrated by concentrated solution of a strong acid or base (meter titrant) to make $pH(2)$ equal to the set point pH in fermenter. E.g., when the organic substrate is neutral, medium 2 is base, and an acid should be used for such a titration. When the organic substrate is an acid, medium 2 is acid, and titration from $pH(2)$ to the set point pH should be made with a base. Then the added amount of the acid or base (per volume to which it is added) will be ΔC_m (mol · l^{-1}). If the molar charge of the meter titrant is T_m, Eq.(9) changes to

$$C_1 - \frac{K_W}{C_1} + \sum_{l_{sub}} C_l(2)\,T_l(C_1) + \Delta C_m T_m = 0 \tag{10}$$

from which the coefficient at $D(2)$ in Eq. (7) equals to $-\Delta C_m T_m$. Then Eq. (7) leads to

$$\mu X = \frac{\Delta C_m |T_m|}{|\lambda_T|} D(2) \tag{11}$$

where

$$\lambda_T = \sum_l \left(\lambda_l + \frac{\delta_l}{\mu}\right) T_l(C_1) - \lambda_q \tag{12}$$

l (summation for substrates and products, except CO_2)

One can easily realize the utilization of moduli considering the sign of T_m and λ_T. E.g. when the organic substrate is neutral, λ_T is close to nitrogen fraction in biomass (see below), $\lambda_T > 0$, the titrant is base, the meter titrant is an acid, from which $-T_m > 0$. Usually coefficients δ_l are connected only with the uptake of substrates or excretion of products linked to energy metabolism. Therefore, when the organic substrate is neutral and organic products are absent (a rather often met case),

$$\lambda_T = \sum_l \lambda_l T_l(C_1) - \lambda_q \tag{13}$$

inorganic substrates

where only ammonia, phosphorous and sulfur substrate ions, metal ions, as well as cell charge are taken into account. We estimated earlier [11] that λ_T in the latter case is close to λ_N, nitrogen fraction in dry cell biomass. However, we noted that λ_q estimation is not sufficiently reliable and demands the additional study.
Thus, for the case of perfect pH maintenance the volumetric rate of cell biomass growth, μX, is proportional to the rate $D(2)$ of medium supply via the pump connected with pH controller:

$$\mu X = AD(2), \quad \text{where} \quad A = \frac{\Delta C_m |T_m|}{|\lambda_T|} \tag{14}$$

This equation relates both to steady-state growth and to transient state in microbial culture. Dimension of A is the same as of X, i.e. gl^{-1}. Dimensions of ΔC_m and λ_T should correspond one to another. In general case ΔC_m and λ_T are measured as moles per liter and moles per g DCB. But when λ_T is close to the fraction of nitrogen in dry biomass and ΔC_m is recalculated to concentration of ammonia nitrogen, it is suitable to measure λ_T as g N per g DCB and ΔC_m as g N per liter.

Dynamic equations and steady state regimes

For the sake of a joint analysis of chemostat and pH-auxostat regimes let us consider the generalized form of culture dynamics equations:

$$\frac{dX}{dt} = \mu(S)X - D(1)\,X - D(2)\,X \tag{15}$$

$$\frac{dS}{dt} = -\frac{1}{Y_{X/S}}\,\mu(S)\,X + D(1)\,S_{01} + D(2)\,S_{02} - DS \tag{16}$$

where S is the residual concentration of the substrate which limits the culture growth in chemostat, g l^{-1}; $Y_{X/S}$ is dry cell biomass yield from the considered substrate, g g^{-1} $\left(\frac{1}{Y_{X/S}} = M_l\left(\lambda_l + \frac{\delta_l}{\mu}\right)\right.$ for the corresponding number; M_l is molar mass of the substrate$\left.\right)$;

S_{01} and S_{02} are concentrations of the substrate in supplied media, g l^{-1}. Using Eq.(14), Eqs. (15) and (16) can be rearranged as follows:

$$\frac{dX}{dt} = \mu(S)\,X - D(1)\,X - \frac{1}{A}\,\mu(S)\,X^2 \tag{17}$$

$$\frac{dS}{dt} = \left(\frac{S_{02}}{A} - \frac{1}{Y_{X/S}}\right)\mu(S)\,X + D(1)\,S_{01} - D(1)\,S - \frac{1}{A}\,\mu(S)\,SX \tag{18}$$

For steady-state values of cell biomass and residual substrate concentrations, $\bar{X}$ and $\bar{S}$, Eqs. (17) and (18) take the following form:

$$\mu(\bar{S})\,\bar{X} - D(1)\,\bar{X} - \frac{1}{A}\,\mu(\bar{S})\,\bar{X}^2 = 0 \tag{19}$$

$$\left(\frac{S_{02}}{A} - \frac{1}{Y_{X/S}}\right)\mu(\bar{S})\,\bar{X} + D(1)\,S_{01} - D(1)\,\bar{S} - \frac{1}{A}\,\mu(\bar{S})\,\bar{S}\bar{X} = 0 \tag{20}$$

Usual theory of chemostat did not take the flow of titrant into account; therefore, it corresponds to case $D(2) = 0$. Account of titrant flow gives rise to $D(2) \neq 0$, $S_{02} = 0$ (see below).
In *pH*-auxostat $D(1) = 0$, $\mu = D(2)$, from which, utilizing Eqs. (19) and (20),

$$\bar{X} = A \tag{21}$$

$$\bar{S} = S_{02} - \frac{A}{Y_{X/S}} \tag{22}$$

Eq. (21) clarifies the microbiological meaning of A: it equals to the value of cell biomass concentration which is obtained in *pH*-auxostat on the medium with the given alkalinity (or acidity, when the organic substrate is an acid). From Eq. (22) it can be seen that A is restricted:

$$A < S_{02} Y_{X/S}, \tag{23}$$

otherwise the amount of the organic substrate would not be sufficient for the culture to change *pH*(2) of the medium to *pH* of the set point, and the flow of medium would be automatically stopped.
Consider the effect of titrant flow on the operation of chemostat: the case $D(2) \neq 0$, $S_{02} = 0$. Interrelations for steady state values of cell biomass and residual substrate concentrations, $\bar{X}$ and $\bar{S}$, in chemostat are obtained from Eqs. (19) and (20):

$$\mu(\bar{S}) = \frac{D(1)}{1 - \dfrac{\bar{X}}{A}} \tag{24}$$

$$\bar{S} = S_{01} - \left(\frac{S_{01} - S_{02}}{A} + \frac{1}{Y_{X/S}}\right)\bar{X} = S_{01} - \frac{C}{A}\,\bar{X} \tag{25}$$

$$\text{where}\quad C = S_{01} - S_{02} + \frac{A}{Y_{X/S}} \tag{26}$$

The limit case for high concentration alkali is: $D(2) \approx 0$, $D(1) \approx D$, $A \to \infty$, from which we obtain the usual equations: $\mu(\bar{S}) = D$, $\bar{S} = S_{01} - \dfrac{1}{Y_{X/S}}\,\bar{X}$.

Dependences of $\bar{X}$ and $\bar{S}$ on the set dilution rate through inlet 1, $D(1)$ (which is a fraction of the total dilution rate) can be found for every specific form of the function $\mu(\bar{S})$. Here we consider the MONOD case which is widely met in practice and suitable for comparison with the classical theory:

$$\mu = \frac{\mu_m S}{K_S + S} \tag{27}$$

Figs. 1 and 2 represent the cases when concentration of alkali in the titrant flow (inlet 2) is correspondingly high and low. We accepted values of culture parameters often met for yeast grown on glucose: $\mu_m = 0.5\ \mathrm{h}^{-1}$, $Y_{X/S} = 0.5$, $K_S = 0.1\ \mathrm{g\ l}^{-1}$[6]. Organic substrate concentrations: $S_{02} = 0$ (i.e. flow 2 serves only as titrant), S_{01} has been calculated for 5 g dry cell biomass per liter when growth is limited by the organic substrate. Fig. 1 is the case of relatively high concentration of NH_3 in titrant solution: 3 per cent w/w. Then $\Delta C_m = 24.7\ \mathrm{g\ N\ l}^{-1}$, $T_m = 1$. λ_1 for yeast is often close to 0.07 g N per g DCB. Therefore, $A = 350\ \mathrm{g\ l}^{-1}$ (Eq. 14). Fig. 2 corresponds to diluted NH_3 solution. Its concentration is selected to make both flows equal when growth is limited by the organic substrate. Then $D(2) = \frac{\mu}{2}$, from which and Eq. (14): $A = 2\bar{X} = 10\ \mathrm{g\ l}^{-1}$. This corresponds to about 0.09% NH_3.

For calculation process it is suitable to accept $\bar{X}$ as an argument and utilize Eqs. (25), (27), as well as

$$D(1) = \mu(\bar{S}) \left(1 - \frac{\bar{X}}{A}\right) \tag{28}$$

$$D(2) = \mu(\bar{S}) \frac{\bar{X}}{A} \tag{29}$$

$$W = \mu(\bar{S})\, \bar{X} \tag{30}$$

where W is the productivity of 1 liter of working volume. After calculation for $\bar{X}_0$ from 0 to $5\ \mathrm{g\ l}^{-1}$ the variables $\bar{X}$, $\bar{S}$, $D(2)$, μ and W were plotted against $D(1)$.

It should be emphasized that $W = \frac{1}{A} D(2)$, i.e. productivity is proportional to the rate of titrant supply.

One can see from Fig. 1 that the concentrated titrant flow is small even in the region of maximal productivity. Therefore, in this case the steady-state curves are essentially the same as classical ones calculated without taking into account the titrant flow [14].

On the contrary, Fig. 2 displays quite a different arrangement of curves. The region of $D(1)$ from zero to μ_m can be clearly divided into two parts which are equal in the given case due to the selected value of A. When the set value of nutrient medium flow $D(1)$ varies from zero to a half of maximum specific growth rate, μ_m, the titrant flow $D(2)$ is close to $D(1)$, and μ increases from zero to almost μ_m. Residual substrate concentration, $\bar{S}$, is small compared with S_{01}. Cell biomass concentration, $\bar{X}$, is practically constant and close to $D(1)\, S_{01} Y_{X/S}/(D(1) + D(2))$ (see Eq. (16)). Productivity increases in proportion to $D(1)$. This band of cultivation regimes is a chemostat region.

When $D(1)$ increases further from a half of μ_m to μ_m, another situation arises. Cell biomass concentration falls to zero and the residual substrate concentration enlarges to S_{01}, both linearly with $D(1)$ increment. Specific growth rate is close to its maximum, μ_m. Therefore the total dilution rate, $D(1) + D(2)$, is approximately constant and $D(2)$ as well as productivity W diminish as $D(1)$ increases. The fall of both $D(2)$ and W is due to the decrement of cell biomass concentration. It is the presence of pH-controlled flow of diluted alkali solution which gives rise to such a difference of Fig. 2 plots from

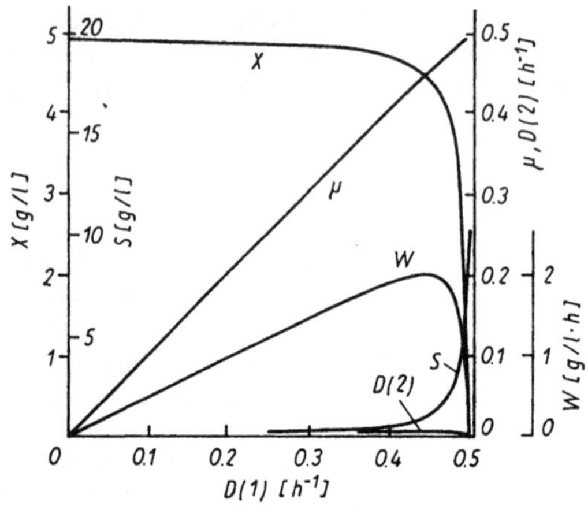

Fig. 1. Steady state culture (chemostat) variables against the set medium flow via inlet 1 (high concentration of alkali in titrant flow).
Abscissa: medium flow via inlet 1, per liter of working volume, D(1).
Ordinates: $\bar{X}$ and $\bar{S}$, concentrations of dry cell biomass and residual substrate; D(2), titrant flow per liter of working volume; $\mu = D = D(1) + D(2)$, specific growth rate; W, productivity of 1 liter of working volume.
Parameters: $\mu_m = 0.5\ h^{-1}$, $K_S = 0.1\ g\ l^{-1}$, $Y_{X/S} = 0.5$, $S_{01} = 10\ g\ l^{-1}$, $S_{02} = 0$, A = 350 (see text).

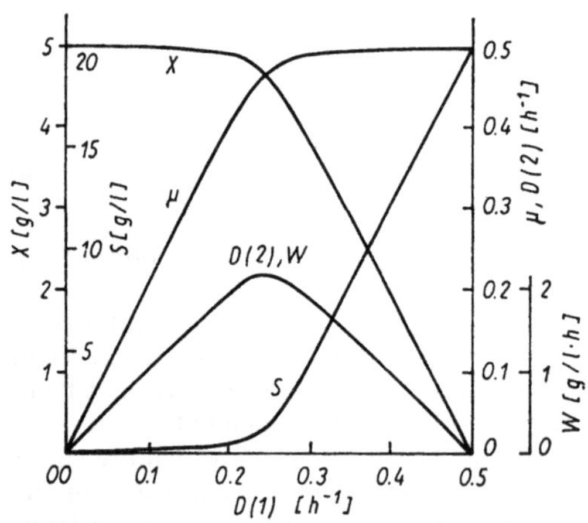

Fig. 2. Steady state culture variables against the set medium flow via inlet 1 (low concentration of alkali in titrant flow).
Abscissa and *ordinates* see legend to Fig. 1.
Parameters: $\mu_m = 0.5\ h^{-1}$, $K_S = 0.1\ g\ l^{-1}$, $Y_{X/S} = 0.5$, $S_{01} = 20\ g\ l^{-1}$, $S_{02} = 0$, A = 10 (see text).

those of Fig. 1. It results in the shift of maximum of W into the middle of $D(1)$ range and lengthening of the parts of Fig. 1 curves adjacent to its right edge. This band of $D(1)$ can be identified as turbidostat or pH-auxostat region. The middle of $D(1)$ range represents the transition from chemostat to pH-auxostat which was usually not realized.

Study of stability

The first method of LYAPUNOV is utilized here. Let

$$X = \bar{X} + \xi \tag{31}$$

$$S = \bar{S} + \sigma \tag{32}$$

where ξ and σ are small deviations of the concentrations X and S from their steady-state values. Let us develop the right-hand sides of Eqs. (17) and (18) as a series in powers of ξ and σ, and restrict our consideration to zeroth- and first-order terms. The sum of zeroth-order terms equals to zero due to the steady state interrelations Eqs. (19) and (20). For small deviations:

$$\frac{d\xi}{dt} = \alpha_{11}\xi + \alpha_{12}\sigma \tag{33}$$

$$\frac{d\sigma}{dt} = \alpha_{21}\xi + \alpha_{22}\sigma \tag{34}$$

where

$$\alpha_{11} = \left(1 - \frac{2\bar{X}}{A}\right)\mu(\bar{S}) - D(1) \tag{35}$$

$$\alpha_{12} = \bar{X}\left(1 - \frac{\bar{X}}{A}\right)\frac{d\mu}{d\bar{S}} \tag{36}$$

$$\alpha_{21} = \left(\frac{S_{02}}{A} - \frac{1}{Y_{X/S}} - \frac{\bar{S}}{A}\right)\mu(\bar{S}) \tag{37}$$

$$\alpha_{22} = \left(\frac{S_{02}}{A} - \frac{1}{Y_{X/S}} - \frac{\bar{S}}{A}\right)\bar{X}\frac{d\mu}{d\bar{S}} - \frac{\bar{X}}{A}\mu(\bar{S}) - D(1) \tag{38}$$

The characteristic equation of this system is

$$\varkappa^2 - (\alpha_{11} + \alpha_{22})\varkappa + (\alpha_{11}\alpha_{22} - \alpha_{12}\alpha_{21}) = 0 \tag{39}$$

where $\varkappa$ is a parameter of a partial solution of Eqs. (33) and (34), viz., $e^{\varkappa t}$. From Eqs. (24) and (25):

$$\bar{X} = A\left(1 - \frac{D(1)}{\mu(\bar{S})}\right) \tag{40}$$

$$\bar{S} = S_{01} - C\left(1 - \frac{D(1)}{\mu(\bar{S})}\right) \tag{41}$$

From Eqs. (35) — (38) and (40) — (41):

$$\alpha_{11} + \alpha_{22} = -CD(1)\frac{\mu(\bar{S}) - D(1)}{[\mu(\bar{S})]^2}\frac{d\mu}{d\bar{S}} - 2\mu(\bar{S}) + D(1) \tag{42}$$

$$\alpha_{11}\alpha_{22} - \alpha_{12}\alpha_{21} = [\mu(\bar{S}) - D(1)]\left[\frac{CD(1)}{\mu(\bar{S})}\frac{d\mu}{d\bar{S}} + \mu(\bar{S})\right] \tag{43}$$

Substitution of Eqs. (26), (42), (43) into Eq. (39) and solving of the square equation yield the following eigenvalues:

$$\varkappa_1 = -\mu(\bar{S}) \tag{44}$$

$$\varkappa_2 = -[\mu(\bar{S}) - D(1)]\left\{\frac{CD(1)}{[\mu(\bar{S})]^2}\frac{d\mu}{d\bar{S}} + 1\right\} \tag{45}$$

This is a general solution which makes it possible to study the stability of a steady state for each specific dependence of μ on the substrate concentration. $\varkappa_1$ is similar to the solution reported earlier [1, 2, 5, 6]. The equation for $\varkappa_2$ is specific for the case of presence of titrant flow. Reciprocals of $\varkappa_1$ and $\varkappa_2$ are characteristic times of decay of deviations from a steady state.

The first eigenvalue, $\varkappa_1$, is always negative and, thus, never serves as a reason of instability. The corresponding characteristic time, $1/\varkappa_1$, becomes longer as μ decreases, which is usually observed experimentally. The second eigenvalue, $\varkappa_2$, is of more interest, and we discuss it here in details.

Let us consider first the case of chemostat. When titrant is strong (high concentration of alkali or acid), A is high (see above). In chemostat, where $\frac{d\mu}{d\bar{S}}$ is not small, the first term in figure brackets in Eq. (45) is much more than 1. From Eqs. (24) and (26):

$$[\mu(\bar{S}) - D(1)]\,C = \frac{\bar{X}}{1 - \frac{\bar{X}}{A}}\,D(1)\left(\frac{S_{01} - S_{02}}{A} + \frac{1}{Y_{X/S}}\right)$$

which tends to $D(1)\,\bar{X}/Y_{X/S}$ as $A \to \infty$. Simultaneously $\mu(\bar{S})$ becomes equal to $D(1)$ since $D(2) \to 0$. Therefore $\varkappa_2$ becomes equal to

$$\varkappa_2 = -(S_{01} - \bar{S})\frac{d\mu}{d\bar{S}} \tag{46}$$

irrespective of S_{02} (since $D(2) \to 0$ and supply of the substrate via inlet 2 becomes negligible even when $S_{02} \neq 0$). For Monod case (Eq. (27)):

$$\varkappa_2 = -\frac{S_{01}[\mu_m - D(1)] - K_S D(1)}{K_S \mu_m}[\mu_m - D(1)]$$

which is similar to that found earlier [1]. At high values of the residual substrate concentration, $\bar{S}$, the derivative of Eq. (27) $\frac{d\mu}{d\bar{S}} \to 0$, $\varkappa_2 \to 0$ which means that steady state stability ceases. This is also the well-known situation in chemostat near wash-out. Substrate inhibition (e.g. when $\mu(S) = \mu_m S/(K_S + S + S^2/K_i)$) makes $\frac{d\mu}{d\bar{S}}$ positive at high $\bar{S}$ which even increases the instability [5, 6].

Here are some important values of $\varkappa_2$ obtained from Eq. (45) for a general case of *pH*-controlled chemostat. In the wash-out point $\bar{X} = 0$, from which (see Eqs. (24) and (25)): $\mu(\bar{S}) = D(1)$, $\bar{S} = S_{01}$. Since $\mu(S_{01}) \neq 0$, $\varkappa_2 = 0$ in the wash-out point, i.e. the culture looses stability completely.

On the other hand, when $D(1)$ is small (theoretically, $D(1) \to 0$), $\bar{S}$ is very small as well. Then, from Eqs. (24), (25): $\bar{X} \approx \frac{S_{01}A}{C}$, $\mu(\bar{S}) \approx \frac{D(1)}{1 - \frac{S_{01}}{C}}$. At low growth rates μ is usually approximately proportional to S, i.e. $\frac{d\mu}{d\bar{S}}$ tends to a certain constant, equal to μ_m/K_S in MONOD case. Then, from Eq. (45):

$$\varkappa_2\Big|_{D(1)=0} = -S_{01}\left(1 - \frac{S_{01}}{C}\right)\frac{d\mu}{d\bar{S}}\Big|_{\bar{S}=0} \tag{47}$$

At large A-value Eq. (47) coincides with Eq. (46). The value of $\varkappa_2$ at low dilution rates has rather a high modules (see below). E.g., in MONOD case at large A, $\varkappa_2 = -\frac{S_{01}}{K_S}\mu_m$, and $|\varkappa_2| \gg \mu_m$ since usually $K_S \ll S_{01}$: saturation constant is small compared with the supplied substrate concentration [6].

Finally, the value of $\varkappa_2$ in the point of maximum of fermenter productivity, $W = \mu(\bar{S})\,\bar{X}$, is of peculiar interest. In this point:

$$\frac{dW}{dD(1)} = \frac{d\mu}{d\bar{S}}\frac{d\bar{S}}{dD(1)}\bar{X} + \mu\frac{d\bar{X}}{dD(1)} = 0 \tag{48}$$

From Eq. (25):

$$\frac{d\bar{S}}{dD(1)} = -\frac{C}{A}\frac{d\bar{X}}{dD(1)} \tag{49}$$

Using Eqs. (48), (49) and (24):

$$C\frac{d\mu}{d\bar{S}} = \frac{\mu^2}{\mu - D(1)} \tag{50}$$

Substitution of Eq. (50) into Eq. (45) indicates that at maximum productivity of fermenter $\varkappa_2 = -\mu(\bar{S})$, where $\bar{S}$ is the corresponding value of substrate concentration. Thus, in this point both eigenvalues become equal.

Figs. 3 and 4 illustrate the above said for case of MONOD dependence of μ on S. It should be noted that in the point of maximum productivity the steady state is still sufficiently

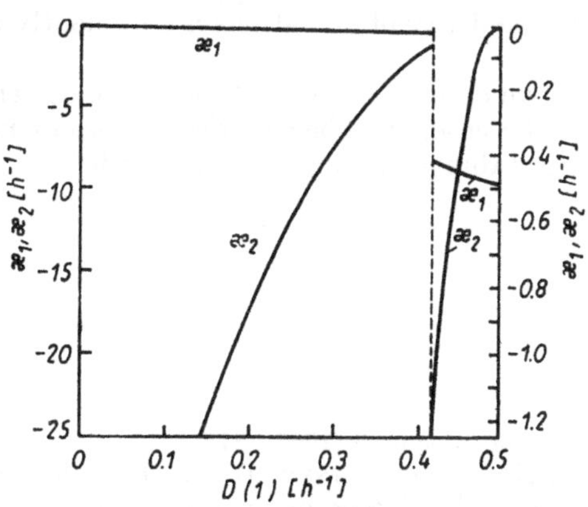

Fig. 3. Eigenvalues of steady state establishment transients in chemostat, $\varkappa_1$ and $\varkappa_2$ (high concentration of alkali in titrant flow).
Abscissa: medium flow via inlet 1, per liter of working volume, D(1).
Ordinates: $\varkappa_1$ and $\varkappa_2$. Since $\varkappa_2$ changes sharply, two different scales of $\varkappa$'s are used; the broken line is the border between the scales.
Parameters: $\mu_m = 0.5\ h^{-1}$, $K_S = 0.1\ g\ l^{-1}$, $Y_{X/S} = 0.5$, $S_{01} = 10\ g\ l^{-1}$, $S_{02} = 0$, A = 350.

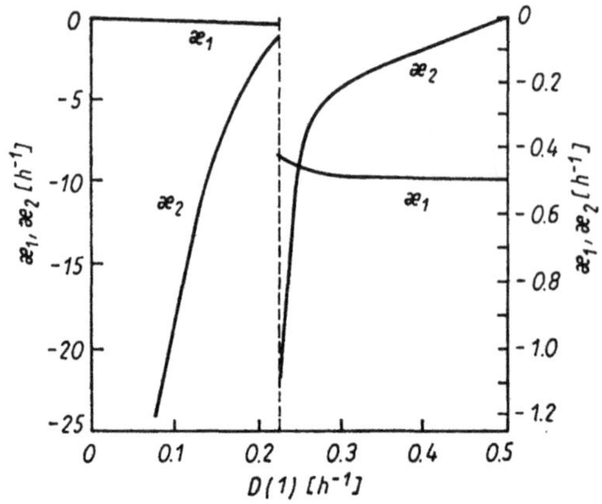

Fig. 4. Eigenvalues of steady state establishment transients in chemostat, $\varkappa_1$ and $\varkappa_2$ (low concentration of alkali in titrant flow). *Abscissa* and *ordinate* see legend to Fig. 3. *Parameters*: $\mu_m = 0.5\ h^{-1}$, $K_S = 0.1\ g\ l^{-1}$, $Y_{X/S} = 0.5$, $S_{01} = 20\ g\ l^{-1}$, $S_{02} = 0$, $A = 10$

stable in any case. But when titrant is concentrated (Figs. 1 and 3), the sensitivity of cultivation regime near this point becomes very strong, and small increase of $D(1)$ leads to a very sharp fall of cell biomass concentration and the stability factor, $|\varkappa_2|$. For diluted titrant this fall is not so sharp.

Another case is *pH*-auxostat: $D(1) = 0$, $\varkappa_2 = -\mu(\bar{S})$ (see Eq. (45)). The presence of two equal eigenvalues originates from the structure of Eqs. (33) and (34) in this case. Utilizing Eqs. (40), (41), (35) –(38) at $D(1) = 0$, we obtain: $\alpha_{11} = -\mu(\bar{S})$, $\alpha_{12} = \alpha_{21} = 0$, $\alpha_{22} = -\mu(\bar{S})$. The system Eqs. (33), (34) becomes uncoupled, both equations having equal coefficients. In pH-auxostat the slope of μ dependence on substrate concentrations does not play any role. This regime is always stable, having a single characteristic time of deviation decay equal to $1/\mu$ (which is close to $1/\mu_m$ when substrate inhibition is absent or small). Correspondingly, the decay of deviations is sufficiently rapid in *pH*-auxostat.

It should be emphasized that the above described relates to the case of perfect stirring and perfect *pH* control. If it is not the case, the characteristic times of mass transfer and *pH* controller will be the culture stability characteristics as well.

Received November 13, 1984

References

[1] Koga, S., Humphrey, A. E.: Biotechnol. Bioeng. **9** (1967), 375.

[2] Romanovsky, Yu. M., Stepanova, N. V., Chernavsky, D. S.: Chto takoe matematicheskaya biofizika (What is the mathematical biophysics, in Russian). Moscow: Prosveshchenie, 1971.

[3] Bergter, F.: Wachstum von Mikroorganismen. Experimente und Modelle. Jena: VEB Gustav Fischer Verlag, 1872.

[4] Fan, L. T., Shah, P. S., Pereira, N. C., Erickson, L. E.: Water Research **7** (1973), 1609.

[5] Romanovsky, Yu. M., Stepanova, N. V., Chernavsky, D. S.: Matematicheskoe modelirovanie v biofizike (Mathematical modelling in biophysics, in Russian). Moscow: Nauka, 1975.

[6] Minkevich, I. G., Utkina, L. I.: Biotechnol. Bioeng. **21** (1979), 357.

[7] MIZOBUCHI, T., MORITA, S., YANO, T.: J. Ferment. Technol. **58** (1980), 33.
[8] PECHURKIN, N. S., POZMOGOVA, I. N., TERSKOV, I. A.: Prikladnaya biokhimiya i mikrobiologiya (Applied Biochemistry and Microbiology, in Russian) **5** (1969), 158.
[9] MARTIN, G. A., HEMPFLING, W. P.: Arch. Microbiol. **107** (1976), 41.
[10] MINKEVICH, I. G.: Biofizika **24** (1979), 712.
[11] MINKEVICH, I. G., EROSHIN, V. K.: Biofizika **24** (1979), 879.
[12] MINKEVICH, I. G.: Biotechnol. Bioeng. **25** (1983), 1267.
[13] MINKEVICH, I. G., NEUBERT, M.: Acta Biotechnol. **4** (1984) 4, 313.
[14] PIRT, S. J.: Principles of Microbe and Cell Cultivation. Oxford, London, Edinburgh, Melbourne: Blackwell Scientific Publications, 1975.

Acta Biotechnol. 6 (1986) 2, 188

Book Review

W. J. YANG

Flow Visualization III

Proceedings of the Third International Symposium on Flow Visualization, September 6–9, 1983
University of Michigan, Ann Arbor, Michigan, USA

Washington, New York, London: Hemisphere Publishing Corporation
Heidelberg, New York, Tokyo, Berlin: Springer-Verlag, 1985, 888 pp., 910 fig., 288 DM.

Since Osborn REYNOLDS' early investigations on the transition from laminar to turbulent flow (Trans. Roy. Soc. London, vol. **174, 1883**) flow visualization has been an important tool in experimental flow mechanics.
The proceedings of the Third International Symposium on Flow Visualization offer on more than 800 pages a vast amount of experimental material both on methods and applications. The methods range from the well-known Schlieren-methods for the visualization of refractive index gradients to interferometric methods and speckle photography, from the insertion of small particles or bubbles in the flow for visualization to light scattering investigations. It is interesting to notice to what large extent lasers are employed in the investigations. This holds e.g. for the widely used light sheet technique and of course for holographic methods. A lot of papers on digital image processing and computer-generated graphics pertinent to flow visualization show "a revolution in the art and science of flow visualization", as the editor stated in the preface of the proceedings. Most of the papers deal with applications. Main topics are: Separated flows, Vortices and Wakes, Jets, Stratified and Boundary Layer-Flows, Rheology, Heat transfer, Biomedical applications and others. The list of possible applications of flow visualization methods might be continued with respect to biotechnology: fermenter design, flow cytometry to mention only a few.
These proceedings will be very helpful for scientists and engineers in many fields of technology. They show the state of art in flow visualization.
On the other hand I would like to remark that there is still a long distance to go to the physical approaches to understand turbulence (see e.g. KADANOFF, Roads to chaos, Physics Today **36** (83), 46) will influence the practically relevant studies of turbulent flow as shown in the papers of the Third International Symposium on Flow Visualization. The proceedings should be recommended to all libraries of instituts and establishments where flow mechanics is of interest, also to scientists looking for applications of lasers and holography.

G. BRAUN

Acta Biotechnol. 6 (1986) 2, 189—199

Mechanismen des Wassertransportes bei der Getreideweiche

MOHR, K.-H.

Ingenieurhochschule für Chemie Köthen
DDR-4370 Köthen, Bernburger Straße 52—57

Summary

A simple method for the investigation of the water transport velocity in cereal kernel is described. The mechanism of the water transport is discussed. It is explained that data, place and strength of the effect of hydrolytic enzymes can be determined on the basis of the changed transport resistance from the temporal change of the different diffusion coefficient. The fundamentals of the modelling of steeping processes are given.

Zur optimalen Führung von Getreideweichen, insbesondere zur Wahl des geeigneten Weichverfahrens für die Produktion eines qualitativ hochwertigen Malzes bei minimalem Wassereinsatz, ist die Kenntnis der im Getreidekorn ablaufenden Mikroprozesse erforderlich. Hierbei nimmt das Vegetationswasser eine Schlüsselstellung ein. Für den Beginn der Lebensprozesse im Getreidekorn ist eine Mindestwassermenge erforderlich. Bei steigendem Energiestoffwechsel hat demgegenüber die ausreichende Sauerstoffversorgung des keimenden Getreides Vorrang. Somit sind Art, Zeitpunkt und Länge der Wasserweichen und Luftrasten genau aufeinander abzustimmen. Für das gewählte Verfahren — Umpumpweiche, Flutweiche, Sprühweiche — sind auf der Grundlage der im Getreidekorn ablaufenden Lebensprozesse die günstigsten Verfahrensparameter zu wählen. Hierzu ist die Kenntnis der effektiven Transportkoeffizienten des Wassers im Getreidekorn in Abhängigkeit von der Weichzeit erforderlich. Aufgrund der Schichtstruktur des Getreides und der mit der Weiche einsetzenden Lebensprozesse im Keimling, die mit einem Strukturabbau des Aleuron und Endosperm verbunden sind, eignen sich zur Bestimmung der effektiven Transportkoeffizienten des Wassers im Getreidekorn nur Verfahren, die die Lebensprozesse des keimenden Getreides nicht beeinflussen. Zu dieser Problematik gibt es in der Literatur keine Angaben.
In Auswertung von Weichversuchen mit bespelzter Gerste mittels 1. FICK'schen Gesetz für stationäre Bedingungen wurde festgestellt, daß die Transportgeschwindigkeit des Wassers in der Reihenfolge Spelte, Aleuron und Endosperm sinkt. Durch den beginnenden strukturellen Abbau des Getreidekorns durch hydrolytische Enzyme ist danach ein erneuter Anstieg festzustellen [1]. Zur exakten Bestimmung der Transportkoeffizienten des Wassers im Getreidekorn beim instationären Prozeß der Getreideweiche ist es erforderlich, ein Auswerteverfahren auf der Grundlage des 2. FICK'schen Gesetzes zu entwickeln.
Ziel vorliegenden Beitrages ist es, ein Verfahren zur Bestimmung der effektiven Diffu-

sionskoeffizienten bei der Getreideweiche vorzustellen. Die mathematischen und experimentellen Grundlagen werden erläutert. Anhand von ausgewählten Beispielen werden Ergebnisse vorgestellt und es werden Ableitungen für technische Weichprozesse gegeben.

Mathematische Grundlagen

Der Wassertransport im Getreidekorn stellt einen molekularen Diffusionsprozeß dar, der für die instationären Bedingungen der Getreideweiche mit dem 2. Fick'schen Gesetz beschrieben werden kann:

$$\partial c/\partial t = D \cdot \nabla^2 \cdot c \tag{1}$$

c — Konzentration des Wassers im Getreidekorn in kg/m³
D — Diffusionskoeffizient des Wassers in m²/s
t — Weichzeit in s
∇ — Nabla-Operator in m⁻¹

Technische Weichprozesse sind dadurch gekennzeichnet, daß der Anfangszustand des Getreides mit der Wasserkonzentration c_A zum Zeitpunkt $t_A = 0$ durch Änderung des Umgebungszustandes bei Weichbeginn gestört wird. Die Konzentrationsfelder $c(r, t)$ mit der Ortskoordinate r in m sind dann zu jedem Zeitpunkt $t > 0$ eindeutig bestimmt, wenn der Umgebungszustand mit der Sättigungskonzentration des Wassers an der Kornoberfläche c_0 für alle Zeiten $t > 0$ als Randbedingung vorgegeben wird. Unter den Bedingungen der Getreideweiche gilt für die Wasseraufnahme die Stoffübergangsbeziehung:

$$j_m = -\beta(c_0 - c_A) \tag{2}$$

j_m — Stoffstromdichte in kg/m² · s
β — Stoffübergangskoeffizient in m/s

Unter Weichbedingungen ist der äußere Stoffübergangswiderstand $1/\beta$ wesentlich kleiner als der innere Stoffleitwiderstand R/D. In diesem Fall wird die Oberflächenkonzentration c_0 selbst eine gegebene Größe:

$$c_0 = c_A(r, t) \tag{3}$$

Unter den Bedingungen der Gl. 3 herrscht Randbedingung 1. Art. Für geometrisch einfache Körper erhält man für den eindimensionalen Stoffstrom bei einmaliger Änderung des Umgebungszustandes von c_A auf c_0 leicht anwendbare analytische Lösungen für Gl. 1. Für die praktische Rechnung ist es vorteilhaft, das instationäre Konzentrationsfeld in Form einer unendlichen Reihe darzustellen, deren Summanden in Abhängigkeit von der Weichzeit mehr oder weniger schnell konvergieren. In dimensionsloser Schreibweise lautet diese Beziehung:

$$\bar{C} = \sum_{n=1}^{\infty} A_n \cdot C_n \cdot e^{-m_n^2 \cdot Fi} \tag{4}$$

Es bedeuten:

$\bar{C} \equiv (c_0 - \bar{c}_t)/(c_0 - c_A)$ die mittlere normierte Konzentration (5)

$Fi \equiv D \cdot t/R^2$ die dimensionslose Zeit (FICK-Zahl) (6)

A_n — Von der Form des ursprünglichen Konzentrationsfeldes abhängige Konstanten

C_n — Koordinatenfunktionen, die die räumliche Konzentrationsverteilung kennzeichnen

m_n — Von den Abmessungen, der Form und den physikalischen Eigenschaften des Körpers abhängige Konstanten

Für den eindimensionalen Stoffstrom in einer Platte, einem Zylinder und einer Kugel für Bi → ∞ sind die ersten 3 Summanden der Gl. 4 in Tabelle 1 zusammengestellt. Die allgemeinen Funktionen können der Literatur entnommen werden [2, 3]. Unter den Bedingungen auch der technischen Getreideweiche ist Bi → ∞ gültig [4, 5]. Die Biot-Zahl *Bi* stellt das Verhältnis des inneren Stoffleitwiderstandes R/D zum äußeren Stoffübergangswiderstand $1/\beta$ dar:

$$Bi \equiv \beta \cdot R/D \tag{7}$$

Tab. 1. Lösungen der Funktionen A_n, C_n und m_n für die ersten 3 Summanden der Gleichung 4 für den eindimensionalen Stoffstrom bei Randbedingungen 1. Art und für $Bi \to \infty$.

Summand	Funktion			Körper
	A_n	C_n	m_n	
1	$4/\pi$	$2/\pi$	$\pi/2$	
2	$-4/3\pi$	$-2/3\pi$	$3\pi/2$	unendliche Platte
3	$4/5\pi$	$2/5\pi$	$5\pi/2$	
1	3,854	0,432	2,405	
2	−5,882	−0,123	5,52	unendlicher Zylinder
3	7,326	0,063	8,653	
1	2	$3/\pi^2$	π	
2	−2	$-3/4\pi^2$	2π	Kugel
3	2	$3/9\pi^2$	3π	

Der prozentuale Fehler in Abhängigkeit von der dimensionslosen Weichzeit für 1···6 Summanden der Reihe (Gl. 4) für die normierte mittlere Konzentration $\overline{C}$ in einem unendlich ausgedehnten Zylinder ist in Abb. 1 grafisch dargestellt. Die integrale mittlere Konzentration des Wassers $\bar{c}_t$ im Getreidekorn nach der Weichzeit t ist entsprechend Gl. 8 definiert:

$$\bar{c}_t = \int_V c \cdot dV \Big/ \int_V dV \tag{8}$$

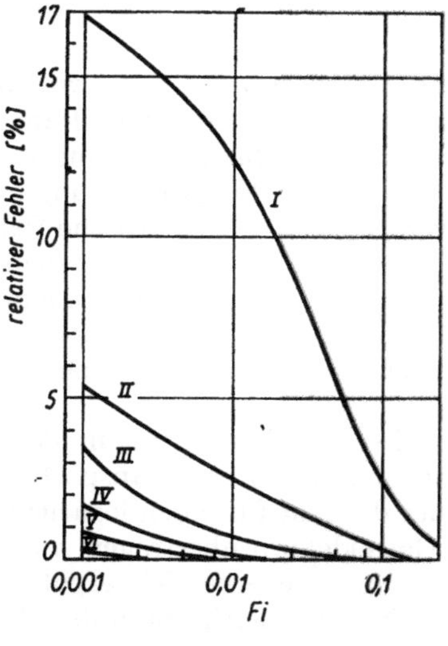

Abb. 1. Abhängigkeit des relativen Fehlers in % von der dimensionslosen Weichzeit (FICK-Zahl) und der Anzahl der Summanden der Gl. 4 für einen unendlich ausgedehnten Zylinder.

Die Diffusionskoeffizienten des Wassers im Getreidekorn betragen $D = 10^{-12} \ldots 10^{-10}\,\mathrm{m^2/s}$. Somit sind die dimensionslosen Weichzeiten auch nach längerer Getreideweiche im allgemeinen kleiner als 0,1. Für genaue Berechnungen ist die Anzahl der erforderlichen Glieder der Gl. 4 in Abhängigkeit von der Weichzeit in Verbindung mit Abb. 1 zu ermitteln. Beispielsweise betragen die gemittelten Diffusionskoeffizienten des Wassers für die Getreideweiche $D_m \approx 10^{-11}\,\mathrm{m^2/s}$ bei einer mittleren Länge des Transportweges $R \approx 1{,}5 \cdot 10^{-3}\,\mathrm{m}$ [6, 7]. Unter diesen Bedingungen ist für eine einstündige Weichzeit $Fi = 0{,}016$ und für eine eintägige Weichzeit $Fi = 0{,}38$. Liegt bei Verwendung nur des 1. Summanden der Gl. 4 der Fehler im ersten Fall bei 12%, beträgt er für den zweiten Fall nur noch 1%. Für kurze Weichzeiten sind daher mindestens drei Summanden der konvergenten Reihe zu verwenden.

Mit Gl. 4 in Verbindung mit Abb. 1 und Tabelle 1 kann für Formkörper bei Kenntnis der Diffusionskoeffizienten die zeitliche mittlere Konzentrationsänderung rechnerisch verfolgt werden. Zur Auslegung von Prozessen ist es auch möglich, auf der Grundlage obiger Gleichung Nomogramme zu entwickeln. Hierzu ist eine gesonderte Publikation vorgesehen [9]. Desgleichen kann Gl. 4 auch rechentechnisch aufbereitet werden.

Analytische Lösungen für mehrdimensionale Konzentrationsfelder können multiplikativ aus eindimensionalen Lösungen zusammengesetzt werden [10]. Beispielsweise erhält man die Lösung für einen endlichen Zylinder zu:

$$\overline{C}(r, x, t) = \overline{C}_1(r, t) \cdot \overline{C}_2(x, t) \tag{8a}$$

wobei $\overline{C}_1$ für den Zylinder und $\overline{C}_2$ für die Platte mit Gl. 4 zu bestimmen sind.

Die experimentelle Bestimmung der Diffusionskoeffizienten des Wassers im Getreidekorn erfolgte auf der Grundlage der Gleichung 4. Diese kann vereinfacht in dimensionierter Form geschrieben werden:

$$(c_0 - \bar{c}_t) = A' \cdot C' \cdot \mathrm{e}^{-m^2 \cdot t} \tag{9}$$

Durch Logarithmieren erhält man:

$$\ln(c_0 - \bar{c}_t) = -m^2 \cdot t + \ln A' \cdot C' \tag{10}$$

Wird diese Gleichung differenziert, ergibt sich:

$$\partial \ln (c_0 - \bar{c}_t)/\partial t = -m^2 \tag{11}$$

Der Faktor m^2 kann leicht experimentell ermittelt werden. Zuerst wird die Anfangskonzentration des Getreides c_A durch Trocknen bei 105°C ermittelt. Danach wird dem Getreide eine Störung durch die beginnende Weiche als Randbedingung 1. Art vorgegeben. Das Getreidekorn erhält dabei zur Zeit $t = 0$ einen Eingangssprung $(c_0 - c_A)$. Für alle Zeiten $t > 0$ lautet dann die Sprungantwort $(c_0 - \bar{c}_t)$. Die grafische Darstellung der Sprungantwort in der Form $\ln (c_0 - \bar{c}_t) = f(t)$ ergibt eine Gerade mit dem Anstieg m^2. In Abb. 2 sind diese Bedingungen grafisch dargestellt. Für geometrisch einfache Körper kann aus dem Anstieg m^2 der Diffusionskoeffizient D berechnet werden.

Die wirklichen zeitlichen und örtlichen Konzentrationsprofile in einer Platte sind in dimensionsloser Form in Abb. 3 dargestellt.

Berechnungsgrundlagen für die Sterilisation fester Lebensmittel in der Verpackung wurden von Ball [8] entwickelt. Für den instationären Wärmetransport durch Leitung im Lebensmittel führte er einen f-Wert ein, der der Zeit für einen logarithmischen Zyklus entspricht. Auf den Wassertransport im Getreidekorn übertragen würde er die Zeit darstellen, in der der Quotient $(c_0 - \bar{c}_t)/(c_0 - c_A)$ um eine Zehnerpotenz vergrößert wird. Die Beziehungen zur Berechnung der Diffusionskoeffizienten des Wassers lauten

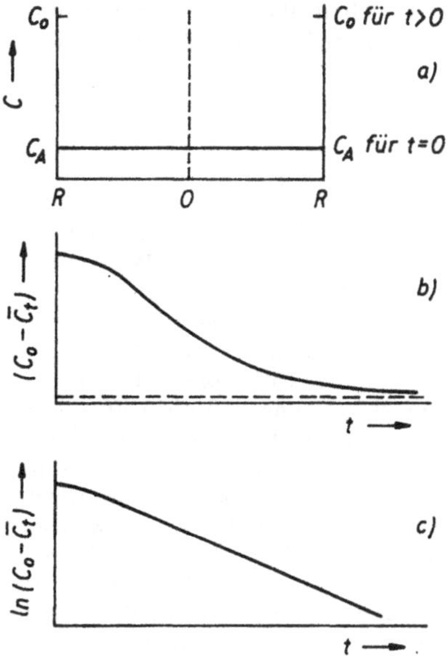

Abb. 2. Zur experimentellen Bestimmung der Diffusionskoeffizienten wird
a) als Randbedingung 1. Art eine Störung als Eingangssprung ($c_0 - c_A$) vorgegeben.
b) Die Sprungantwort ($c_0 - \bar{c}_t$) $= f(t)$ wird experimentell bestimmt und
c) als Anstieg der Geraden bei der grafischen Darstellung von $\ln(c_0 - \bar{c}_t) = f(t)$ ermittelt.

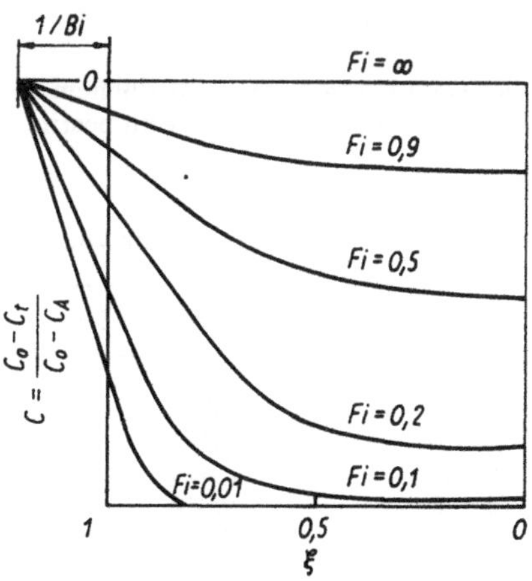

Abb. 3. Konzentrationsfeld C (ξ, Fi, Bi) einer Platte bei Randbedingung 1. Art.

Korrektur: $c = \frac{c_0 - \bar{c}_t}{c_0 - c_A}$

dann:

— Endlicher Zylinder: $$D = \frac{0{,}398}{\left[\frac{1}{R^2} + \frac{0{,}427}{(0{,}5\,L)^2}\right] f} \quad (12)$$

— Kugel: $$D = 0{,}233 R^2/f \quad (13)$$

— Quader: $$D = 0{,}233/(1/X^2 + 1/Y^2 + 1/Z^2)\, f \quad (14)$$

Für irregulär geformte Partikel kann auf der Grundlage ihrer Form und Größe in Anlehnung an die Gl. 12, 13 und 14 nur eine Modifizierung der Konstanten akzeptable Ergebnisse bringen.

Experimentelle Grundlagen

Die experimentellen Bedingungen zur Bestimmung der Diffusionskoeffizienten des Wassers im Getreidekorn wurden so gewählt, daß Randbedingung 1. Art herrschte. Weichversuche wurden mit bespelzter Gerste, Weizen, Roggen und geschältem Roggen durchgeführt. Aus der Verschiedenheit des Untersuchungsmaterials sollte ermittelt werden, ob die Spelte, das Schälen oder der unterschiedliche Eiweißgehalt Einfluß auf die Transportkoeffizienten des Wassers haben. Es kann bereits an dieser Stelle gesagt werden, daß der Verlauf der Diffusionskoeffizienten in Abhängigkeit von der Weichzeit für alle Proben prinzipiell gleich war und daß die Diffusionskoeffizienten für eine bestimmte Weichzeit sich nur geringfügig unterschieden. Die Anfangskonzentrationen des Wassers im Getreide c_A wurden durch Trocknen bis zur Massenkonstanz bestimmt. Für die Weichversuche wurden Getreidechargen von ca. 1 kg in mindestens 3 kg thermostatiertes Wasser unter leichtem Rühren geschüttet. Unter diesen Bedingungen war $Bi \to \infty$ gewährleistet. Nach verschiedenen Weichzeiten wurden Proben entnommen und durch Wägen und Trocknen wurde der Wassergehalt bestimmt. Die Probenmenge wurde so gewählt, daß sie mindestens 1000 Körner enthielt. Damit war gewährleistet, daß die Eigenschaften des Kornspektrums der Proben stets die mittleren Eigenschaften eines Kornes repräsentierten. Aus diesen Meßwerten wurde der funktionelle Zusammenhang $\bar{c}_t = f(t)$ berechnet. Die Konzentrationen c_0 wurden durch Weichen bis zur Sättigung erhalten. Zur Bestimmung der f-Werte erfolgte die Darstellung der Meßwerte in der Form

$$\log (c_0 - \bar{c}_t) = f(t) \tag{15}$$

Tabelle 2 enthält als Beispiel einen Weichversuch für Weizen bei einer Weichtemperatur von 18 °C.

Die grafische Darstellung der Meßwerte entsprechend Gl. 15 ergibt einen charakteristischen und für alle untersuchten Getreidearten prinzipiell gleichen Kurvenverlauf. Für das Beispiel der Tabelle 2 ist er in Abb. 4 grafisch dargestellt.

Für eine Weichzeit bis zu 5 Stunden wurden vier Geraden mit unterschiedlichem Anstieg und drei Knickpunkten erhalten. Aus dem Anstieg der Geraden wurden die f-Werte bestimmt, die zur Berechnung der D-Werte erforderlich sind.

Tab. 2. Weichversuch von Weizen bei einer Weichtemperatur von 18 °C

t [s]	y_W	$\bar{c}_t$ [kg/m³]	$\log (c_0 - \bar{c}_t)$
0	0,0720	91,7	2,630
1200	0,2093	227,0	2,461
2400	0,2458	256,4	2,414
3600	0,2757	278,9	2,375
4800	0,2980	294,8	2,345
6000	0,3153	306,7	2,320
7200	0,3288	315,5	2,302
8400	0,3390	322,2	2,287
9600	0,3453	326,4	2,278
10800	0,3506	329,8	2,270
12000	0,3598	335,3	2,257
13200	0,3715	342,7	2,238
14400	0,3840	350,0	2,220
15600	0,3973	357,9	2,198
16800	0,4116	366,2	2,175
18000	0,4273	374,8	2,149

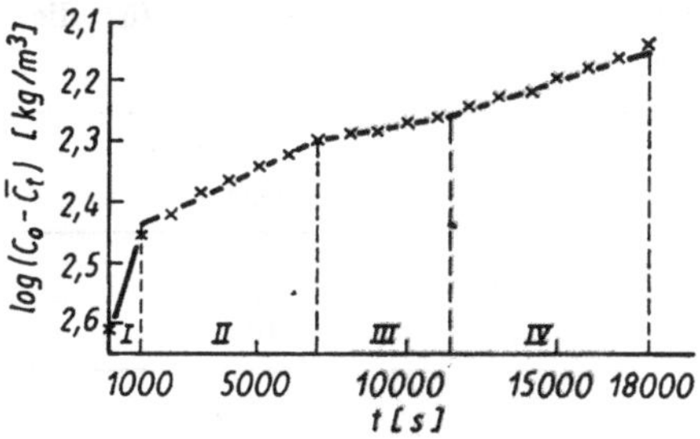

Abb. 4. Grafische Darstellung der Meßwerte der Tabelle 2 in der Form log $(c_0 - \bar{c}_t) = f(t)$.

Im Falle des Getreides hat man es nicht mit geometrisch einfachen Körpern zu tun. Zur Berechnung der D-Werte aus den f-Werten mit den Gl. 12, 13 oder 14 ist es erforderlich, die Eigenschaften eines Kornes durch die mittleren Eigenschaften des Kornspektrums zu charakterisieren. Dazu wurden mit dem Mikrometer die großen Achsen a_i und die kleinen Achsen b_i von 1000 Körnern ausgemessen und die erforderlichen Größen der Klassen i des Spektrums berechnet. Diese Untersuchungen zeigten, daß bereits mit maximal 300 Körnern repräsentative Werte erhalten werden. Die für die weiteren Rechnungen erforderlichen charakteristischen Meßwerte sind für das Beispiel Weizen in Tabelle 3 zusammengestellt. In Anlehnung an Gl. 12 für den endlichen Zylinder wurde zur Berechnung der D-Werte aus den gemessenen f-Werten für die untersuchten Getreidearten folgende modifizierte Gleichung verwendet:

$$D = 0{,}279/[(1/(b_m/2)^2 + 0{,}427/(a_m/2)^2] \cdot f \qquad (16)$$

Tab. 3. Charakteristische Meßwerte für Weizen

Formelzeichen	Dimension	Stoffwerte Weizen
$d_{PS} \cdot 10^3$	m	4,15
$d_{Pm} \cdot 10^3$	m	4,10
$a_m \cdot 10^3$	m	6,12
$b_m \cdot 10^3$	m	3,38
E_m	—	1,81
$A_P \cdot 10^6$	m^2	17,44
A	m^2/kg	0,383
$m_{P,1000} \cdot 10^3$	kg	48,08
ϱ_{G_T}	kg/m^3	1302
ϱ_{m_A}	kg/m^3	1274
ϱ_{m_0}	kg/m^3	1146
y_{W_A}	kg/kg	0,0720
y_{W_0}	kg/kg	0,4500
$c_{.t}$	kg/m^3	91,7
c_0	kg/m^3	515,7

Die mit dieser Beziehung berechneten Diffusionskoeffizienten für die vier verschiedenen Bereiche der Getreideweiche für eine maximale Weichzeit von 5 h sind für das Beispiel Weizen in Tabelle 4 zusammengestellt.

Die Überprüfung der so ermittelten D-Werte erfolgte durch Nachrechnung der örtlichen und zeitlichen Konzentrationsprofile im Getreidekorn für ein Zweischichtenmodell mit der GAUSS'schen Fehlerfunktion [4, 5, 11]. Dazu wurden die Schichtdicken von Aleuron und Endosperm mikroskopisch bestimmt. Die berechneten Weichkurven wurden mit den experimentell bestimmten verglichen. In Abb. 5 sind für das Beispiel Weizen bei einer Weichtemperatur von 18 °C berechnete und gemessene Weichkurve grafisch dargestellt. Die mittlere Abweichung lag für die berechnete Kurve bei +5,1%. Auch bei anderen Weichtemperaturen und Getreidearten lag die Abweichung zwischen berechneter und gemessener Weichkurve innerhalb dieser Grenzen. Somit ist die halbempirische Gl. 16 ausreichend für die Berechnung der D-Werte aus den experimentell bestimmten f-Werten.

Tab. 4. Diffusionskoeffizienten und Aktivierungsenergien für die verschiedenen Bereiche der Getreideweiche für das Beispiel Weizen

Bereich	Temperatur [°C]	f-Wert [s]	$D \times 10^{11}$ [m²/s]	E [kJ/mol]
I	10	7645	9,2	16,6
I	18	6702	10,5	
II	10	71000	1,0	37,0
II	18	42105	1,7	
III	10	380000	0,2	65,0
III	18	175000	0,4	
IV	10	65000	1,1	16,6
IV	18	60000	1,2	

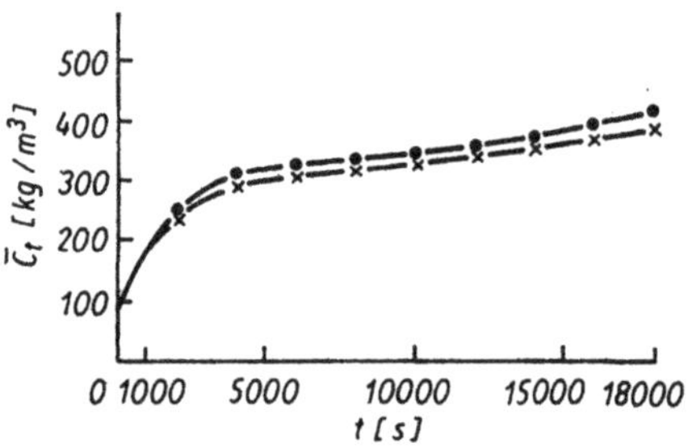

Abb. 5. Vergleich der experimentell bestimmten Weichkurve (×——) mit der mittels Gauß'scher Fehlerfunktion berechneten Weichkurve (●——) für Weizen bei einer Weichtemperatur von 18 °C.

Ergebnisse und Diskussion

Zur Bestimmung der Sättigungskonzentrationen c_0 wurden für alle untersuchten Getreidearten die gesamten Weichkurven aufgenommen. Für das Beispiel bespelzte Gerste ist in Abb. 6 eine derartige Weichkurve grafisch dargestellt. In dieses Bild sind für das gleiche Beispiel die berechneten Diffusionskoeffizienten des Wassers in Abhängigkeit von der Weichzeit für den gesamten Weichbereich eingetragen. Der Abb. 6 ist zu entnehmen, daß der Diffusionskoeffizient des Wassers im Getreidekorn in starkem Maße von der Weichzeit abhängt.

Der Weichbeginn ist durch eine sehr schnelle Wasseraufnahme gekennzeichnet. In diesem Bereich lagen die Diffusionskoeffizienten für alle untersuchten Getreidearten in der

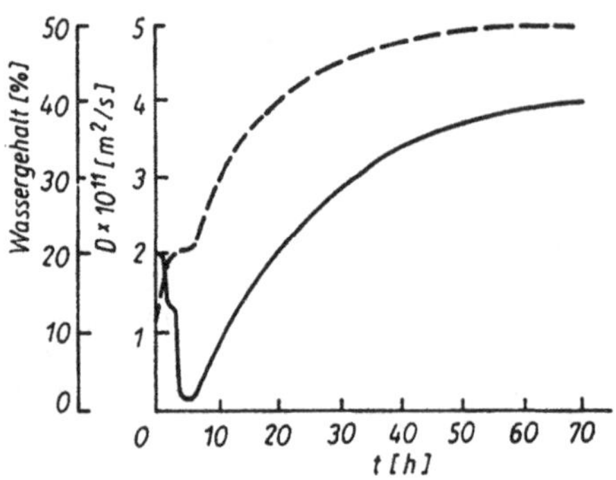

Abb. 6. Abhängigkeit der Wasseraufnahme y_W (unterbrochene Kurve) und des Diffusionskoeffizienten D (ausgezogene Kurve) von der Weichzeit t für Gerste bei 18 °C.

Größenordnung von 10^{-10} m²/s. Die Diffusionskoeffizienten für die nicht behinderte Diffusion des Wassers betragen im Temperaturbereich von 10···33 °C nach Messungen von MILD [12] $D = 1{,}5\cdots2{,}5 \times 10^{-9}$ m²/s. Somit handelt es sich bei der ersten Wasseraufnahme durch das Getreidekorn um eine schwach behinderte Diffusion. Die von LAWRENCE [13] gemessenen Diffusionskoeffizienten des Wassers in rohen Kartoffeln lagen in der gleichen Größenordnung. Die Aktivierungsenergien für die nichtbehinderte Wasserdiffusion lagen nach Messungen von MILD [12] bei 16,6 kJ/mol. Für die erste Wasseraufnahme durch das Getreidekorn wurden gleiche Werte für die Aktivierungsenergie erhalten, wie der Tabelle 4 für das Beispiel Weizen entnommen werden kann. Daraus kann geschlossen werden, daß die Transportgeschwindigkeit des Wassers in diesem Bereich nur gering von der Weichtemperatur beeinflußt wird. Zu Weichbeginn wird das Vegetationswasser durch an der Kornbasis mündende Gefäße, dem sogenannten Aufsaugepithel, aufgesaugt und an den Keim und das Schildchen weiter transportiert. Nach einer Weichzeit von zwei Stunden ist in den einzelnen Kornbestandteilen folgender Wassergehalt vorhanden: Spelte — 40,5%; Keimlingsachse — 33,0%; Schildchen — 18,5%; Aleuron — 11,4%; Endosperm — 10,0% [14]. Bereits ab einem Wassergehalt von 30% beginnen die Lebensäußerungen des Keimlings. Bei einem Wassergehalt des Keimlings von 38% kommt es zu einer raschen Keimung, die mit einer starken Enzymbildung verbunden ist. Durch Streckung der Zellen des Aufsaugepithels wird die Aufnahmefläche für das Vegetationswasser stark vergrößert. Diese lokalisierte Wasseraufnahme ist für die schnelle Bildung der Gibberelline in der Stammanlage und deren Transport zum Keimling und über das Schildchen zum Aleuron erforderlich.
Bereits nach einer Weiche von etwa 20 Minuten sinkt der D-Wert des Wassers auf ein Zehntel ab und liegt im Mittel bei $D \approx 10^{-11}$ m²/s. Nach dieser Zeit ist die sehr schnelle Wasseraufnahme durch das Aufsaugepithel abgeschlossen. Sie erfolgt auf diesem Wege nur noch entsprechend der Transportgeschwindigkeit des Wassers im Keimling. In dieser zweiten Phase der Wasseraufnahme erfolgt der Transport durch die Zellulosewände der Spelte zum Aleuron. Nach einer Weichzeit von etwa 2···3 Stunden ist der Wassergehalt im Aleuron so hoch, daß dort durch die Gibberelline die Endohydrolasen aktiviert werden können.
Danach erfolgt ein weiteres Absinken des Diffusionskoeffizienten auf $D \approx 10^{-12}$ m²/s. Während beim zweiten Teilschritt der Wasseraufnahme durch die Aleuronschicht eine Aktivierungsenergie von ca. 37 kJ/mol erhalten wurde, beträgt sie für den dritten Teilschritt ca. 65 kJ/mol. Daraus kann entnommen werden, daß beim zweiten Teilschritt die Behinderung des diffundierenden Wassers durch die Proteine des Aleurons bewirkt wird. Beim dritten Teilschritt müssen dagegen erst die Eiweißumhüllugen der stärkeführenden Zellen proteolytisch abgebaut werden, ehe der hydrolytische Abbau der Stärke erfolgt.
Nach einer Weichzeit von ca. vier Stunden beginnt der Diffusionskoeffizient des Wassers merklich anzusteigen und erreicht bei Sättigung Werte von etwa $4 \cdot 10^{-11}$ m²/s. Dieser Bereich ist durch einen steigenden enzymatischen Abbau der Struktur des Endosperm gekennzeichnet. Dieser D-Wert entspricht dem von MEDDINGS [15] in geschrotetem Gerstenmalz gemessenen ($D = 3{,}6 \cdot 10^{-11}$ m²/s). Für diesen Bereich wurde eine Aktivierungsenergie von 16,6 kJ/mol erhalten.
Aus den Geschwindigkeiten der ersten Wasseraufnahme durch das Aufsaugepithel, der zweiten Wasseraufnahme durch Spelte und Samenschale, den Zeitpunkt der Bildung der Gibberelline und Endohydrolasen sowie den Strukturabbaus des Endosperms ergeben sich wichtige Rückschlüsse auf die Praxis des Weichens und damit verbunden auf den Wasserbedarf in Mälzereibetrieben.
Moderne Weichverfahren sind durch kurze Wasserweichzeiten bis zu genau festgelegten Weichgraden gekennzeichnet. Während langer Trockenperioden findet der Konzentrationsausgleich im Korn statt. Weichverfahren mit kurzen Wasserweichzeiten und langen

Luftrasten führen zu einer wesentlichen Beschleunigung der Enzymentwicklung, die eine wichtige Voraussetzung für eine Verkürzung der Weich- und Keimzeit ist. Ein Zusatz von Gibberellinen zum ersten Weichwasser kann gleichfalls eine merkliche Verkürzung der Weichzeit bewirken. Eine Verkürzung der Weichzeit durch Vakuum, Druck oder mechanische Beeinflussung des Getreides ist nicht möglich. Die erste Wasserweiche sollte auf maximal vier Stunden ausgedehnt werden. In dieser Zeit ist der Wassergehalt der einzelnen Kornbestandteile wie folgt angestiegen: Keimlingsachse — 43,6%; Spelte — 43,1%; Schildchen — 36,8%; Endosperm — 12,5% [14]. Diese Wassergehalte sind ausreichend, um alle bereits geschilderten enzymatischen Prozesse im Korn ablaufen zu lassen. Während der folgenden etwa zwanzigstündigen Trockenweiche wird die Struktur soweit enzymatisch abgebaut, daß bei der folgenden kurzen Wasserweiche schnell hohe Weichgrade erreicht werden. Bei derartigen Weichen liegt der praktische Wasserbedarf nur um etwa 30···100% über dem theoretischen Bedarf von 0,7 m^3 Wasser/t Gerste.
Soll der Weichprozeß beispielsweise rechnerisch oder nomografisch behandelt werden, erweist es sich ohne große Fehlerabweichungen als vorteilhaft, mit mittleren Diffusionskoeffizienten D_m zu arbeiten. Für die Beispiele bespalzte Gerste und Roggen sind die D_m-Werte für den Temperaturbereich von 10···25 °C als Funktion der reziproken absoluten Temperatur in Abb. 7 grafisch dargestellt. Für praktische Rechnungen können sie dieser Abbildung entnommen werden. Mit den gegebenen Meßwerten ist die mathematische Behandlung des Weichprozesses möglich.

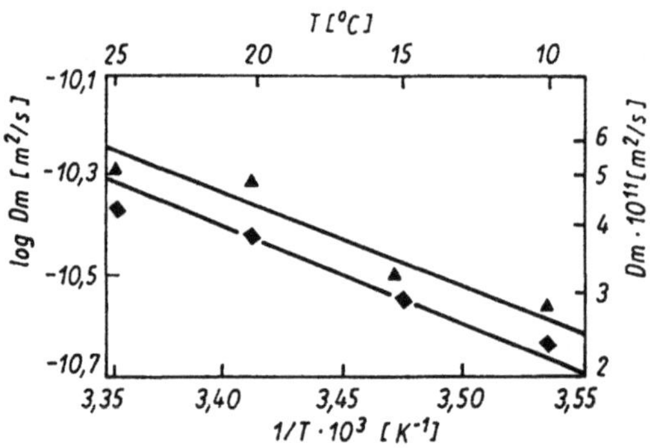

Abb. 7. Mittlere Diffusionskoeffizienten D_m in Abhängigkeit von der Weichtemperatur für Gerste (▲——) und Roggen (◆——).

Nomenklatur

A	Spezifische Oberfläche	m^2/kg
A_p	Oberfläche eines Kornes	m^2
a	Kornlänge	m
Bi	Biot-Zahl	—
b	Korndurchmesser	m
c	Konzentration	kg/m^3
$\bar{c}_t$	Mittlere Konzentration nach der Weichzeit t	kg/m^3
d	Durchmesser	m
D	Diffusionskoeffizient	m^2/s
E	Exzentrizität	—
Fi	Fick-Zahl	—
f	f-Wert	s
m	Masse	kg
N	Kornzahl	—
$m_{p,1000}$	Masse von 1000 Körnern	kg

R	Radius	m
r	Ortskoordinate	m
t	Zeit	s
y	Massenanteil	kg/kg
β	Stoffübergangsgeschwindigkeit	m/s
ϑ	Temperatur	°C
ϱ	Dichte	kg/m³

Indizes

A	Anfang
G	Getreide
i	Klasse
m	mittlere Größe
o	Sättigung
T	Trocken
W	Wasser

Eingegangen: 25. 7. 1984

Literatur

[1] Mohr, K.-H.: Acta Biotechnol. **2** (1982), 297.
[2] Bronstein, I. N., Semendjajew, K. A.: Taschenbuch der Mathematik. Leipzig: Teubner Verlagsgesellschaft B. S. B. G. B., 1977. 17. Auflage.
[3] Verein Deutscher Ingenieure: VDI-Wärmeatlas, 5. Düsseldorf: VDI-Verlag GmbH, 1974. 2. Auflage.
[4] Mohr, K.-H.: Lebensmittelindustrie **31** (1984) 4.
[5] Mohr, K.-H.: Lebensmittelindustrie **31** (1984) 5.
[6] Mohr, K.-H.: Lebensmittelindustrie **28** (1981), 19.
[7] Mohr, K.-H.: Lebensmittelindustrie **29** (1982), 67.
[8] Ball, C. C., Olsen, F. C. W.: Food Research **20** (1955), 666.
[9] Mohr, K.-H.: Lebensmittelindustrie **31** (1984) 6.
[10] Kunis, J., Mohr, K.-H.: Lebensmittelindustrie **31** (1984), 13.
[11] Crank, J.: The Mathematics of Diffusion. Oxford: Clarendon Press, 1975.
[12] Mild, K. H., Lovtrup, S.: Biochim. Biophys. Acta **373** (1974), 383.
[13] Lawrence, J. G., Scott, R. P. W.: Nature **210** (1966) 5033, 301.
[14] Schuster, K., Weinfurter, M., Narziss, L.: Die Bierbrauerei, Bd. 1. Die Technologie der Malzbereitung. Stuttgart: Ferdinand Enke Verlag, 1976. 6. Auflage.
[15] Meddings, P. S., Potter, O. E.: Cereal Chem. **51** (1974), 1.

Acta Biotechnol. 6 (1986) 2, 200

Book Reviews

K. C. Marshall

Microbial Adhesion and Aggregation

Dahlem Workshop Reports
Life Sciences Research Report 31
Berlin, Heidelberg, New York, Tokyo: Springer-Verlag, 1984. 423 S., 86 Abb., 35 Tab., 59 DM

Das Buch erscheint in der Reihe der Dahlem Workshop Reports und berichtet über eine Konferenz, die vom 15.—20. 1. 1984 durchgeführt wurde. Entsprechend dem Charakter dieser Konferenzen handelt es sich um Berichte, die von international angesehenen Fachleuten als multidisziplinäre Übersichtsarbeiten angelegt sind, wobei die auf der Konferenz diskutierten neuen Ergebnisse, Experimente, Konzepte, Methoden und Modelle berücksichtigt werden. Besonders herausgearbeitet sind aktuelle Entwicklungen und Hinweise auf künftige Forschungsrichtungen.
Die Hauptabschnitte sind Mechanismen der Adhäsion (7 Beiträge), Biofilmentwicklung (6 Beiträge), Aktivität an Oberflächen (5 Beiträge) und die Aggregation (5 Beiträge). Der Bogen dieser interdisziplinären Forschung ist weit gespannt: Adhäsion an unbelebten, tierischen und pflanzlichen Oberflächen, genetische Kontrolle, chemische Charakterisierung von Biofilmen, vergleichende Physiologie von anhaftenden und frei lebenden Mikroorganismen, Transformation von Festsubstraten, Flockulation von Hefe als Modellfall einer Aggregation und Bildung von Kolonien und Konsorten von Mikroorganismen. Das Buch ist eine Fundgrube für den Biotechnologen, der sich mit immobilisierten Zellsystemen, Feststoffermentationen oder mikrobieller Korrosion befaßt. Der Mediziner stößt auf Ergebnisse, die das Verständnis von Intestinal- oder Zahnerkrankungen vertiefen. Dem Ingenieur erschließen sich neue Gesichtspunkte für die Effektivierung von Trennprozessen.
Die Terminologie ist für interdisziplinäre Verständigung oft ein schweres Hindernis. Dem wurde vorgebeugt durch ein kleines Glossarium, in dem 20 wichtige Begriffe definiert sind. Auch der umfangreiche (15 Seiten) Sachindex ist für einen Kongreßbericht durchaus keine Selbstverständlichkeit. Das Buch behandelt ein lange vergessenes und nun in schneller Entwicklung begriffenes Kapitel der angewandten Mikrobiologie.
Es kann jedem, der auf diesem Gebiet arbeitet oder lehrt, empfohlen werden.

K. Sattler

E. H. Houwink

A Realistic View on Biotechnology

Frankfurt/M.: DECHEMA — EFB, 1984. 134 pp., 30 DM.

The European Federation of Biotechnology (EFB) has gained momentum since its foundation in 1978 in bringing together experts from universities, research institutes and industry, representing the rapidly developing scientific and technological disciplines which are cooperating in the various fields of biotechnology in Europe.
On the occasion of the 3rd European Congress on Biotechnology in München (September 1984) a professional and realistic view on the status of Biotechnology has been prepared for the press and the interested public.
The Executive and the Science Advisory Committees of the European Federation of Biotechnology have entrusted the presentation of this picture to the author. It cannot be complete and may even raise controversial points. Nevertheless the author, to the best of his knowledge, has given a survey which balances very well the long term visions, the realistic view of the present situation and the steps to be taken to realise the opportunities which are emerging from the new biotechnologies.

Acta Biotechnol. 6 (1986) 2, 201—204

Short Communication

Silierung durch aerobe Säuerung — paradoxer Einfall oder praktische Lösung?

SATTLER, K., BABEL, W.

Akademie der Wissenschaften der DDR
Institut für Biotechnologie, Leipzig
DDR-7050 Leipzig, Permoserstraße 15

Summary

For ensiling of herbage which is rich in protein and therefore difficult to ferment a procedure is proposed. It is based on aerobic acidification using biological deoxydants. The crops are treated with a suspension of living cells of *Acetobacter methanolicus* and a water solution of ethanol. This treatment brings about environmental conditions which promote lactic acid fermentation: anaerobiosis, complex nutritional supply and protective concentrations of H^+-ions. Preliminary attempts have confirmed the principle of conservation by aerobic acidification combined with biological deoxydation in agricultural practice. Damages of legume crops with high protein content are prevented. The advantages of the new ensiling process is discussed in comparison to traditional treatment.

Einleitung

In der DDR sind 50—60% der zur Silierung kommenden Futtermassen schwer vergärbar.
Es handelt sich um junges Gras, Futtergetreide, Rotklee u. a. in einer Menge von etwa 20 Mill. t. Diese Pflanzen sind gekennzeichnet durch einen hohen Eiweiß- und geringen Zuckergehalt [1]. Eine Möglichkeit, die Zuckerkonzentration zu erhöhen und auch bei solchen Grünmassen das Silierergebnis zu verbessern, ist das Anwelken. Diese Maßnahme ist jedoch abhängig von der Witterung und kann nicht verhindern, daß es immer noch bei 30—40% dieses Pflanzenmaterials zu Fehlgärungen kommt. Die jährlichen Verluste betragen 7—8 Mill. t, das sind nahezu 10% des Gesamtaufkommens an Grobfutter. Diese Menge entspricht ca. 25 kt an verdaulichem Rohprotein.
In vielen Ländern Mittel- und Nordeuropas gelten analoge Verhältnisse.
Der Zweck jeder Silierung besteht in der Konservierung von Pflanzenmaterial. Konservierung bedeutet in diesem Fall, daß durch Einschränkung bestimmter mikrobieller Aktivitäten, wie Oxydation (Atmung) und Hydrolyse (Abbau von Makromolekülen) die pflanzliche Substanz nach Zusammensetzung und Futterqualität weitgehend erhalten bleibt, nach Möglichkeit bis zum Beginn der nächsten Vegetationsperiode.

Übliche Verfahrensweisen

Nach den üblichen Verfahrensweisen wird die Konservierung erreicht durch rasches Absenken des pH-Wertes auf Werte um 4 und durch Einstellung anaerober Bedingungen. Anaerobiose wird durch technische Maßnahmen, wie Verdichtung und Abdeckung der Pflanzenmasse [2, 3, 9] und durch Zusatz sauerstoffzehrender Chemikalien, z. B. Natriummetabisulfit [2, 3], erzeugt. Sie ist ein wichtiger Milieufaktor für den intensiven Verlauf der Milchsäuregärung. Komplikationen ergeben sich bei eiweißreichen Pflanzen mit unzureichendem Zuckerangebot und hoher Pufferkapazität u. a. dadurch, daß der pH-Wert in Bereichen bleibt, die Fehlgärungen, z. B. Buttersäurebildung durch Chlostridien, zulassen. Abgesehen von den Nährstoffverlusten sind die Folgen des Verderbs, daß solche Silagen schlecht aufgenommen und die tierischen Produkte in ihrer Qualität beeinträchtigt werden [2]. Eine Konservierung schwersilierbarer Futterstoffe ist auch durch Zusatz von Säuren möglich, bevorzugt Ameisensäure. Ameisensäure ist gegenüber vielen unerwünschten Mikroorganismen bakterizid. Im Unterschied zu Mineralsäuren, die von den Tieren ungern aufgenommen werden und deren Salze den tierischen Stoffwechsel belasten, wird sie im Verdauungstrakt, sogar unter Energiegewinn, abgebaut. Sie beeinträchtigt weder die Futteraufnahme noch die Produktqualität [2, 4].
Über den Einsatz von Ameisensäure entscheidet im wesentlichen ihre Verfügbarkeit. Der Bedarf ergibt sich aus der Menge von schwer vergärbaren Futterstoffen und den notwendigen Zusätzen zum Siliergut (0,2—0,8% vom Frischfuttergewicht).
Es übersteigt bei weitem die Produktion, die nach den Anforderungen der Textil- und Lebensmittelindustrie bemessen ist. Während diese Anwendungsgebiete unumstritten sind, bleiben die Meinungen über die Zweckmäßigkeit des Einsatzes von Ameisensäure als Silierhilfsmittel bis heute geteilt, nicht zuletzt auch im Hinblick auf arbeitsschutztechnische Probleme, die bei der Bevorratung und Ausbringung bestehen [2]. Es wäre ein Vorteil, könnte man für diesen Zweck auf Grundchemikalien zurückgreifen, deren Aufkommen so groß ist, daß der Bedarf als Silierhilfsmittel nur einen geringen Anteil am Gesamtbedarf ausmacht, und die nicht korrosiv und gesundheitsschädlich sind.
Die bloße Konservierung mit Hilfe von Mineralsäuren oder anderen Chemikalien, die bakterizid oder bakteriostatisch wirken, richtet sich letztlich auch gegen die Milchsäurebakterien [2]. Sie ist im Hinblick auf Nebenwirkungen und Aufwendungen bedenklich und sollte besser durch einen komplexen biologischen Prozeß ersetzt werden, der die natürliche Fähigkeit des Pflanzenmaterials zur Milchsäuregärung nützt und begünstigt.
Durch Laborversuche und in ersten Praxisversuchen konnte gezeigt werden, daß auf der Grundlage einer aeroben Säuerung den genannten Anforderungen in einigen Punkten entsprochen und das Problem einer Lösung näher gebracht werden kann.

Grundzüge der Silierung durch aerobe Säuerung

Silierung ist dem Wesen nach eine Konservierung durch *anaerobe Säuerung*. Es mag daher auf den ersten Blick paradox erscheinen, daß gerade dabei eine *aerobe Reaktion* hilfreich sein soll. Nachfolgend wird gezeigt, daß durch die Kombination von aerober Oxidation und anaerober Milchsäurebildung der Prozeß der Silierung intensiviert und stabilisiert wird.
Die aerobe Reaktion ist in unserem Fall die Oxidation von Ethanol zu Essigsäure. Verursacht wird sie durch ein „Essigsäurebakterium" aus der Gattung *Acetobacter* [6—8]. Sie erfolgt unter Verbrauch von Sauerstoff aus der Umgebung. Ethanol und Essigsäurebakterien bilden zusammen ein biologisches Desoxydans, das mit der durchaus erwünschten Nebenwirkung einer Säurebildung ausgestattet ist. Werden „Essigsäurebakterien"

und Ethanol einer frisch geschnittenen Pflanzenmasse zugegeben, die in üblicher Weise mit „Milchsäurebakterien" besetzt ist, so kommt es zur Sauerstoffzehrung im Futterstock bei gleichzeitiger Essigsäurebildung. Sauerstoffzehrung wird im Futterstock schon mit geringen Ethanolmengen (0,5—1 kg pro Tonne Frischfutter) erreicht. Säuerung, die für eine bakteriostatische Wirkung ausreicht, erfordert besonders bei eiweißreichen Pflanzen mit hoher Pufferkapazität dem gegenüber ein Mehrfaches (5—10 kg pro Tonne Frischfutter) an Ethanol.
Im Ergebnis dieser und jeder anderen Ansäuerung kommt es zur Permeabilisierung der Pflanzenzelle, äußerlich erkennbar am Turgorverlust des pflanzlichen Gewebes. Wie

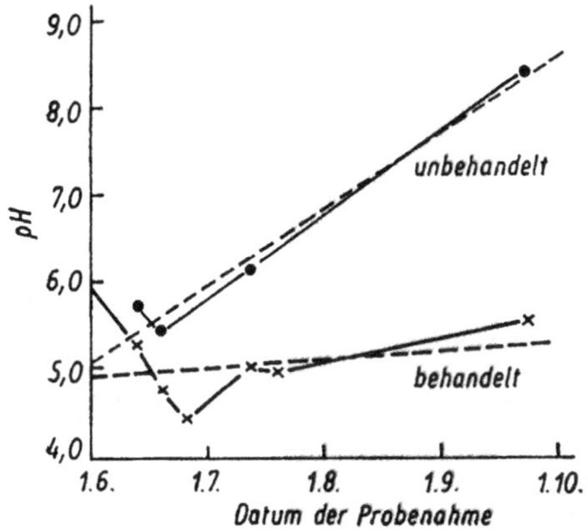

Abb. 1. pH-Wert — Vergleich behandelte und unbehandelte Proben.
•— $y = 5{,}06 + 0{,}029\,x$; ×— $y = 4{,}95 + 0{,}003\,x$
Der Versuch wurde in der landwirtschaftlichen Praxis durchgeführt. Das Silo wurde mit frisch geschnittener Luzerne (625 t) zu einem frühen Zeitpunkt (Ende Mai) und unter ungünstigen Witterungsbedingungen (Dauerregen) angelegt. Die Behandlung erfolgte mit 5 kg Ethanol und 400 g Lebendbakterien (ca. 100 g Bakterientrockensubstanz) pro Tonne Frischfutter

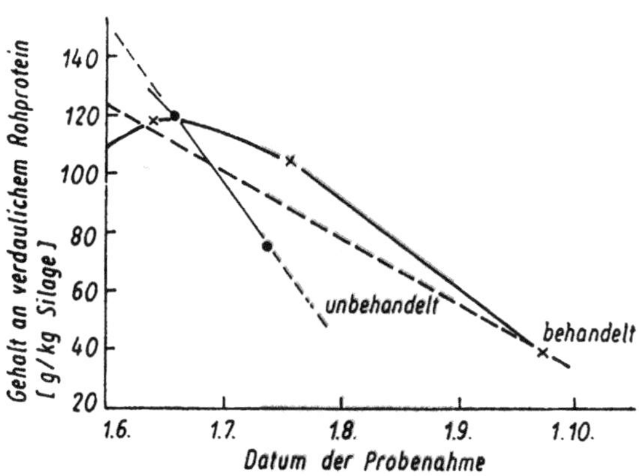

Abb. 2. Gehalt an verdaulichem Rohprotein (g/kg Silage).
Vergleich behandelte und unbehandelte Proben.
•— $y = 150{,}8 - 1{,}863\,x$; ×— $y = 122{,}7 - 0{,}750\,x$
Versuchsbedingungen wie bei Abb. 1.

beim Welken ist dieser Vorgang begleitet von der Freisetzung von Zellinhaltsstoffen, wie z. B. von Zuckern, Vitaminen und Aminosäuren. Durch eine Maßnahme werden somit alle die Entwicklung von Milchsäurebakterien begünstigenden Faktoren geliefert: Anaerobiose, komplexes Nährmedium und protektive H^+-Ionenkonzentration.
In dieser komplexen Wirkung der aeroben Säuerung besteht der wesentliche Unterschied und der Vorteil gegenüber dem einfachen Zusatz von Säuren. Eiweißverluste, wie sie für alle Säurekonservierungen von eiweißreichen Pflanzen typisch sind [5], werden allerdings auch damit nicht vermieden (Abb. 1 u. 2). Durch Behandlung nach dem Prinzip der aeroben Säuerung oder mit biologischen Desoxydantien kann zunächst nur erreicht werden, daß Silagen aus eiweißreichen Pflanzen nicht verderben, sondern als Futtermittel erhalten bleiben.

Eingegangen: 19. 3. 1985

Literatur

[1] Wojahn, E., Weise, G.: Wiss. Fortschr. 33 (1983) 7, 250.
[2] Schmidt, W., Wetterau, H., Beyrich, H., Müller, M., Gottschling, E. M.: Silageherstellung. Berlin: VEB Deutscher Landwirtschaftsverlag, 1973.
[3] McDonald, P.: The Biochemistry of Silage. J. Wiley & Sons, 1981.
[4] Weissbach, F., Hein, E., Schmidt, L.: XIII. Int. Graslandkongreß, Leipzig 1977, 171.
[5] Demarquilly, C., Grenet, E., Dulphy, J.-P.: Int. Graslandkongreß, Leipzig 1977, 200.
[6] Steudel, A., Miethe, D., Babel, W.: Z. Allg. Mikrobiol. 20 (1980) 10, 663.
[7] Babel, W.: Acta Biotechnol. 4 (1984) 4, 389.
[8] Sattler, K., Wünsche, L., Engel, J., Babel, W.: Molekulare Genetik, Mikrobiologie und Virologie (Moskau) 1 (1985), 33.
[9] Weise, G., Rettig, H., Suckow, G.: XIII. Int. Graslandkongreß, Leipzig 1977, 269.

Acta Biotechnol. 6 (1986) 2, 205

Book Reviews

H. G. Gassen, A. Martin, S. Bertram

Gentechnik — Einführung in Prinzipien und Methoden

Uni-Taschenbücher 1290
Stuttgart: Gustav Fischer Verlag, 1985. 362 S., 178 Abb., 33 Tab., 28,80 DM

Die Autoren versuchen mit diesem allgemeinverständlichen Buch sowohl Interesse für die Gentechnik zu erwecken als auch die zum Teil recht verworrenen Vorstellungen von den Möglichkeiten und Zielen dieses sich stürmisch entwickelnden Fachgebietes in realistischere Bahnen zu lenken. Es ist sicher nicht leicht, das Verhältnis von Darstellung der für das Verständnis erforderlichen Grundlagen und dem eigentlichen Sachverhalt zu wahren und es dennoch einem möglichst großen Leserkreis nahezubringen. Diese Schwierigkeiten haben die Autoren gut bewältigt.
Obwohl die 16 Kapitel von 8 verschiedenen Autoren stammen, sind sie doch in Stil und Verständlichkeit gut durchdacht und recht einheitlich. Die Kapitel sind didaktisch sehr gut aufgebaut und werden durch einfache und verständliche Abbildungen ergänzt. Lediglich das Kapitel zur Morphologie und Funktion der Zelle scheint den Rahmen zu sprengen. Dafür hätte man sich eine breitere Diskussion zu denkbaren Anwendungsmöglichkeiten vorstellen können. Ebenso sollte bei einer Überarbeitung das Stichwortverzeichnis erweitert werden, um das Verständnis der Fachausdrücke noch mehr zu erleichtern. Trotzdem kann man dieses Buch jedem an der Genetik und Gentechnik Interessierten empfehlen.

E. Spanier

A. Laitha

Handbook of Neurochemistry
Vol. 2. Experimental Neurochemistry

Second Edition. New York, London: Plenum Press, 1982. 475 pp., $ 59.50

Der Herausgeber, A. Laitha, hat das Gebiet der Experimentellen Neurochemie in 19 ausgewählten Kapiteln von erfahrenen Autoren darstellen lassen. Auffällig ist das gelungene Bemühen, den Stoff so aktuell darzubieten, daß der Leser den Trend der Entwicklung dieses sich schnell ausweitenden Gebietes vermittelt bekommt. Alle Kapitel sind in gleicher Weise sorgfältig gestaltet und hinreichend mit Bildern, Diagrammen und Tabellen versehen und thematisch in sich geschlossen. Die über 2500 Literaturzitate erleichtern darüberhinaus den Zugriff zu Originalarbeiten, Übersichtsartikeln und Handbüchern. Neben vier spezifisch neurologisch angelegten Kapiteln „Diagnosis of Hereditary Neurological Metabolic Diseases“, „Human Brain Postmortem Studies of Neurotransmitters and Related Markers“, „Neurological Mutants“ und „Perfusion of the Isolated Brain“ sind die fünfzehn weiteren methodischen Kapitel zwar mit neurologischen Bezug abgefaßt, aber anregend und gewinnbringend auch für Wissenschaftler anderer Disziplinen zu lesen, die sich mit zellulären und subzellulären Fragestellungen befassen. Das Kapitel „RNA-DNA-Hybridization: Analysis of Gene Expression“ tangiert zum Beispiel taxonomische und phylogenetische Fragen der Prokaryoten.
Die Kapitel „Receptor Mapping by Histochemistry“, „Receptor Measurement“, „Rapid Enzyme Inactivation“, „Radioenzymatic Analysis“, „Two-Dimensional Polyacrylamide Gel Electrophoresis of Proteins“, „The Identification of Subcellular Fractions of the CNS“, „Cell Isolation“, „Principles of Compartmentation“, „Analytical Aspects of the Pharmacokinetics of Psychotrophic Drugs“, „Principles and Application of PET in Neuroscience“, „Selected Micromethods for Use in Neurochemistry“, „Mass Spectrometric Analysis of some Neurotransmitters and her precursors and Metabolites“, „Gas Chromatography“ and „High Performance Liquid Chromatography“ empfehlen sich von selbst.
Dieses Handbuch wird jedem einschlägig interessierten Wissenschaftler im gleichen Sinne nützlich sein, wie es Volker Neuhoff in diesem Buch seinen Fachkollegen rät: „Neurochemists seeking a specific micromethod should look to other disciplines ...“.

A. Schmidt

Acta Biotechnol. 6 (1986) 2, 206

Book Review

N. Blin, M. F. Trendelenburg und E. R. Schmidt

Molekular- und Zellbiologie

Berlin, Heidelberg, New York, Tokyo: Springer-Verlag, 1985. 199 S., 32 DM

Die Herausgeber legen mit diesem Buch eine Themensammlung zu aktuellen Gebieten der Molekular- und Zellbiologie vor. Die vor allem methodisch orientierten Beiträge veranschaulichen allgemeinverständlich die gegenwärtigen Möglichkeiten der molekular- und zellbiologischen Forschung. Die einzelnen Beiträge sind so geschrieben und instruktiv illustriert, daß der biologisch interessierte Nichtfachmann einen guten Überblick über das gegenwärtige methodische Repertoire des Molekularbiologen erhalten kann. Der Aktualität der Themen, weniger der Vollständigkeit wurde dabei der Vorrang gegeben.

Themen zu den drei Schwerpunkten

— Charakterisierung spezieller Proteine
— DNA-Analyse
— Funktionsuntersuchungen an Genen und Chromosomen

stellten die Herausgeber in den Mittelpunkt. Das Buch stellt eine zwar nicht umfassende, aber sehr nützliche Einführung zu diesem Wissenschaftsgebiet dar. Es ermöglicht eine gute Vororientierung und erleichtert die Erschließung der sehr vielfältigen Fachliteratur.
Die Abfassung des Buches in deutscher Sprache kommt den Lesern im deutschsprachigen Raum entgegen und eröffnet einem breiten Leserkreis Einblicke in den modernsten Stand dieser Forschungsrichtung.
Die Herausgeber konzipierten das Buch vor allem für Studenten höherer Semester, die sich einen Einblick in eine Reihe moderner Aspekte des weitumfassenden Gesamtkomplexes Zell- und Molekularbiologie verschaffen wollen. Darüber hinaus ist es allen molekularbiologisch interessierten Nichtfachleuten als eine methodische Einführung mit Überblickscharakter zu empfehlen.

A. Steudel

Acta Biotechnol. 6 (1986) 2, 3 U.-Seite

Book Review

H. U. BERGMEYER (Editor-in-Chief)
J. BERGMEYER, M. GRASSL

Methods of Enzymatic Analysis (3rd Edition)

Vol. VII. Metabolites 2: Tri- and Dicarboxylic Acids, Purines, Pyrimidines and Derivatives, Coenzymes, Inorganic Compounds

Weinheim: VCH Verlagsgesellschaft, 1985. 641 S., 325 DM (Einzelpreis), 285 DM (bei Abnahme aller Bände)

In diesem 7. Band der neuen Bergmeyer-Auflage (5 Abschnitte, 58 Kapitel) wird, wie bereits im 6. Band, ein Gebiet behandelt, das maßgeblich zur Entwicklung der enzymatischen Analyse beigetragen hat. Neben bekannten Methoden, die auf den neuesten Stand gebracht wurden, werden zahlreiche moderne enzymatische Meßverfahren vorgestellt, die neben einer genaueren Metabolitbestimmung in biologischen Materialien vor allem auch eine solche mit wesentlich höherer Empfindlichkeit (bis zum Femtomolbereich) erlauben (z. B. Adenosinphosphate, Nicotinamid- bzw. Flavin-Coenzyme, CoA, Enzymimmunoassay bei cAMP). Hervorzuheben ist weiterhin, daß man auch in diesem Band Bestimmungsmethoden für eine ganze Reihe von Metaboliten bzw. anorganischen Substanzen findet, die bisher im „Bergmeyer" noch nicht enthalten waren (D-Malat, D-Tartrat, Panthethein, Diadenosintetraphosphat, S-Adenosylmethion und -homocystein, Tetrahydrofolat, UDP-2-desoxyglucose und -galactose, UDP-glucosamin und -galactosamin, 5-Fluoro-UDP-glucose und -galactose, Hydrogencarbonat, Sulfit, Mg, Zn, Pb, Arsenat). Andererseits sind im Vergleich zur 2. Auflage (1974) für einige Substanzen keine Bestimmungsmethoden mehr enthalten, nicht zuletzt deshalb, weil diese Substanzen als Metabolite in freier Form in biologischen Materialien nicht oder nur in sehr geringer Konzentration auftreten. Ebenso verzichtet der Herausgeber in dieser 3. Auflage bewußt auf die Bestimmungsmethoden für Nukleinsäuren und Polyribonukleotide.

Im 1. Abschnitt werden 13 Methoden zur Bestimmung von 10 Di- bzw. Tricarbonsäuren vorgestellt (Maleat und meso-Tartrat nicht mehr behandelt). Der 2. Abschnitt bringt 9 Methoden für 9 verschiedene Purine, Pyrimidine bzw. Nukleoside; außerdem eine Methode zur Bestimmung des Gesamtgehaltes an Purinen und nicht-phosphorylierten Purinnukleosiden. Methoden für Cytosin, Cytidin, Desoxycytidin, -thymidin und -uridin sind nicht mehr enthalten. Im umfangreichsten, dritten Abschnitt sind 54 Methoden für 36 verschiedene Nukleotide, Coenzyme und verwandte Verbindungen zusammengefaßt. Acrylyl-, Benzoyl-, Crotonyl-, 3-Hydroxypropionyl-3-Hydroxybutyryl- und Malonylsemialdehyd-CoA, Coenzym B 12 sowie die Bestimmung aller Adenosinphosphate mittels alkal. Phosphatase plus Adenosindeaminase wurden nicht mehr aufgenommen. Der 4. Abschnitt bringt 8 Methoden für 12 Nukleosid-diphosphat-zucker und der 5. Abschnitt 9 Methoden für 9 anorganische Substanzen sowie einen Hinweis mit Verweis auf Band IV zur Bestimmung von Blei.

R. BERGER

Acta Biotechnologica

Volume 6 | 1986 | Number 2

Contents

Short Communication

Acta Biotechnologica is indexed in Current Contents/ET & AT;
Chemical Abstracts; Biotechnology Abstracts; Excerpta Medica

Bei Fragen zur Produktsicherheit wenden Sie sich bitte an:
If you have any questions regarding product safety,
please contact:

Walter de Gruyter GmbH
Genthiner Straße 13
10785 Berlin
productsafety@degruyterbrill.com